Collaborative Research
and Development Projects

Tom Harris

Collaborative Research and Development Projects

A Practical Guide

Foreword by Ray Browne

With 16 Figures

 Springer

Dr Tom Harris
www.hi-consulting.com

There are no warranties, expressed or implied, on any of the methods or
approaches described in this book. You must use them at your own risk
and seek legal and professional advice as appropriate. They do however
represent my best endeavors to convey good practice in creating and
running a successful collaborative research and development project.

Library of Congress Control Number: 2007923126

ISBN 978-3-540-46052-7 Springer Berlin Heidelberg New York

Springer is a part of Springer Science+Business Media

springer.com

© Springer-Verlag Berlin Heidelberg 2007

Production: LE-TEX Jelonek, Schmidt & Vöckler GbR, Leipzig
Cover-design: WMX Design GmbH, Heidelberg

SPIN 11884750   42/3100YL - 5 4 3 2 1 0   Printed on acid-free paper

Foreword

By Ray Browne

Developing new techniques, implementing innovative technology and solving awkward problems have always been keen interests of mine. I spent the first twenty years of my career undertaking scientific research and the last twenty-five years helping industry do the same, largely through supporting government sponsored research and development programmes. The scientists, engineers and business people who undertake such R&D projects are all great people for whom I have considerable respect. They are usually very enthusiastic and manifestly pragmatic, but they still need as much help as they can get, particularly in attracting the support that is often essential if they are to realise the full benefits of their work. I am sure this book will provide them with this support in a truly practical and user-friendly manner.

Collaborative R&D, especially when external funding is involved, is not only about meeting deadlines and deliverables but perhaps more importantly about ensuring a return on the investment made in the project. This book's particular strength lies in its focusing attention on achieving this return. It also shows a crucial dependence on sound project management, not only of the core R&D but also of the commercial activities that are all pre-requisites for success.

The author, Tom Harris, is a friend and colleague who I have known as a scientist, successful business man and consultant. In this book he combines his considerable experience with practical hands-on advice that will be of benefit to anyone involved in collaborative projects. This is not another theory-based textbook, it is essential reading and I recommend it to all aspiring collaborators.

Ray Browne is currently a Deputy Director in the Office of Science and Innovation at the Department of Trade and Industry in the UK where he is responsible for managing the department's funded research and development projects.

Acknowledgements

There are many people that deserve my thanks for the inspiration and creation of this book. The approaches and methods have been developed through my experiences working with many projects as an academic, industrialist and monitoring officer, together with the adaptation of project management techniques developed by, among others, J Rodney Turner of Henley Management College.

From my academic and industrial days, I would particularly like to thank Prof John MacIntyre of Sunderland University, Prof Ian Nabney of Aston University, Paul Wilkinson, Chris Kirkham and Lee Gamlyn, TWI, Arjo Wiggins, SP Tyres and Cardionetics Ltd.

I would also like to thank all those that are involved with the UK Technology Programme. Special thanks go to Ray Browne and his team at DTI, especially Lee Vousden and Jackie Whalley; David Crawford and the team at TUV NEL; the UK Research Councils especially Emily Nott of EPSRC and Jim Clipson of NERC; Jon Tenner of Execulence; Helen Lucas; Audrey Canning; and the projects, assessors and monitoring officers I work with, especially Graham Hesketh of Rolls Royce, Gillian Arnold of IBM, John Kelly of DKR Electrical, Paul Donachy of Queens University Belfast, Martin Dove of Cambridge University, and Neil Lewis of Enigma Interactive.

Of all these individuals, Ray Browne deserves my special thanks for his friendship and support and for honouring me by writing the foreword for this book.

Last, but by no means least, I thank my wonderful wife, and once again, not just for doing all the diagrams.

Contents

Introduction

Getting involved in a collaborative research and development project can be one of the most exciting, rewarding and business changing experiences you and your organisation ever take on. The opportunities to push the boundaries of technology, solve major problems, open up new markets and be part of a leading edge team, drive thousands of companies and universities to work together every year. All economies recognise these benefits and encourage these collaborations by offering grant funding, tax relief schemes and other benefits worth hundreds of millions of dollars each year.

For many projects the outcomes are highly beneficial, including the development of new products, services and the opening up of new market opportunities. However for too many, the project does not bring all that was hoped for. The reasons behind these lost opportunities are varied but a great deal of responsibility lies in poor appreciation of what is involved in running a collaborative project and in poor planning and communication between the partners. Perhaps the most common and serious reason however, is the lack of appropriate understanding of how to exploit the results of the project.

This book has been written to provide you with a user guide for your collaborative R&D project. We will start with an analysis of grant funding, looking to see what help might be available to you and how best to secure it. We will look at the legal arrangements that need to be entered into between you and your partners. In the third chapter we get much more hands on and I provide methods for planning your project and building a sound business case for it. Next we look at the special project management considerations involved in collaborative projects and present methods to keep the momentum going.

The following two chapters are provided to help you understand the other members of your consortium. If you are an industrial or commercial animal and have not worked with academics before, this short guide will provide an insight into their motives and working practices. The same is provided for academics to help you appreciate the commercial world so that you can understand what drives your industrial partners.

In the next chapter we look at what can go wrong and how best to firstly recognise and avoid the problems, and secondly how to deal with them if you find yourselves in trouble.

The penultimate chapter deals with exploitation. Although the last but one chapter in the book, it is perhaps the most important and describes a process that starts before the project and continues long after.

Finally we look at what happens at the end of the project and how your management approach changes. We also look at evaluating the results of your hard work.

My own career started as a university researcher developing novel computing systems for analysing complex data. As I built my research group, it was the applications that were more exciting that the technology itself and so all the projects that we took on were collaborative. We worked with the healthcare, energy generation, aerospace and paper sectors to name but a few. After spinning out a company from one of these projects, I was then involved in collaborative R&D projects from the industrial side. Currently my management consultancy, Hi Consulting, is part of the delivery team for the UK Department of Trade and Industry's Technology Programme. This programme supports in excess of $1.5 billion worth of collaborative R&D. My role in the programme is practical as well as strategic, briefing the applicants, assessors and monitoring officers. I am also responsible for the project monitoring regime adopted by the programme to monitor all its investments.

This book has grown from the experiences on all three sides of the triangle, academic partner, industrial partner and finally funding organisation and project monitor. It is intended to help you get the most from your collaborative project by understanding what will be involved, how to run it professionally and most importantly, how to exploit the results to the full.

So, let's get down to business.

1 Grant funding

Whilst not every collaborative research project is supported by funding from outside the consortium, a large percentage of projects are eligible for support from the governments, charities and other organisations. Not surprisingly, all these organisations lay down criteria for what they can and can't support. The rules get stricter when tax payers' money is involved and where the support must comply with national and international regulations.

This chapter aims to provide a general understanding of grant funding rules and an explanation of the types of things grant assessors look for in making funding decisions. It is obviously not possible to cover every individual grant scheme here and so you must thoroughly check the criteria and rules for the schemes you wish to apply for.

1.1 States aids

The rules for funding projects with tax payers' money vary greatly between the US and Europe. European companies are able to receive funding for commercially focused projects and retain the intellectual property that is generated. The funding is increasingly targeted on projects intended to give European companies a competitive advantage in global markets. In the United States the philosophy tends much more towards funding fundamental research. Another key difference is that the intellectual property developed through US state funded projects does not become the property of the participants; the IPR is placed in the public domain for all the market competitors to share, even overseas competitors.

These different approaches form part of the high profile dispute between US and European aircraft manufacturers about the levels of state assistance each are receiving in order to compete with the other.

European states aid rules

Although the funding available in Europe is more commercially focused, there are still strict regulations that control state funding of research and development programmes. The regulations are laid down in the European treaties. A secretariat in Brussels provides clearance to individual state programmes and monitors that the regulations are properly adhered to. It is worth understanding these regulations since they may help you focus your application for funding. Also if an error is made, you may have to pay the grant back.

The basic principal is that state assistance which could distort competition and affect trade by favouring certain undertakings or producing certain goods, is against the interests of the common market and is therefore prohibited. Fortunately however, there are a few exceptions to this rule. Member states are allowed to assist in disaster recovery, operate the common agricultural policy and support public transport systems for example. Of more interest to us though, is that they are allowed *to facilitate development of certain economic activities or of certain economic areas where this will not adversely affect trading conditions*. In other words, member states are allowed to support companies working in economically depressed regions or working on strategically important technologies or activities.

Most member states take advantage of these exceptions to offer a range of support measures to their businesses. In addition the European Union itself runs Framework programmes that fund projects that span Europe.

What can be funded?

The specific funding rules will vary from scheme to scheme depending on what the funding body wants to achieve and who they are trying to attract. You will have to check out the exact criteria of each scheme as you apply to them. There are however some common principles that are generally followed.

In many cases small and medium sizes enterprises (SMEs) are treated particularly well. This is because it is good to encourage small companies to innovate and grow; they create employment and new opportunities in the market place. There are many grants available to individual companies for investigating innovative ideas, building prototypes and carrying out market research. Here though, we are more interested in the role of SMEs within collaborations. In many cases the involvement of SMEs is keenly encouraged, even to the extent that they can receive extra funding.

Am I an SME?

According to 2006 European definitions, a small company is one that:

- has fewer than 50 employees

- has either a turnover of less than €10 million or a balance sheet total of less than €10 million

- is independent, in other words is not owned by more than 25% by a medium or large enterprise

A medium sized company is one that:

- has fewer than 250 employees

- has either a turnover of less than €50 million or a balance sheet total of less than €43 million

- is independent, in other words is not owned more than 25% by a large enterprise

---

In the United States there is no standard definition for a small business. Generally it is determined by the industry in which it competes, where income and number of employees is used to determine whether a company is a small business or not. Many government contracts are 'set aside' or limited to small businesses only, most often involving services or minor construction.

To be eligible for grant funding, the research will often need to be pre-competitive. This means that the outcomes or results of the research will not be in a state where they can be immediately commercially exploited. Further development will be required before this can take place. The more additional work that will be required, the further from market the project is said to be.

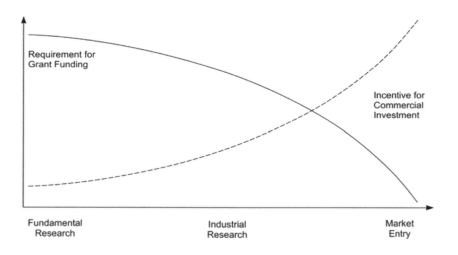

Fig. 1.1. Requirement for grant funding decreases as near-marketness increases

The level of grant that can be provided often depends on this near-market measure. Fundamental research where projects develop results that are very far from exploitation, are often funded at between 75 and 100%. Industrial research, those which could be exploited after say 3 years of further development are normally funded at 50%. Near market or pre-competitive development work which might be exploitable after say 1 year is often restricted to just 25% funding.

These limits make sense when you consider that states use these grants to incentivise companies to carry out risky research that they would otherwise not have the confidence to fully invest in. The more risky the research, the further from realising a return, the more incentive has to be provided to make it happen.

1.2 Additionality

The first thing a grant awarding body will want to know is why you need their money. Why don't you fund the work yourself? This is a perfectly fair question, especially where tax payers' money is involved.

The question is often called the 'additionality' question because basically the funding body wants to know what added value is created by them funding your project.

There are a wide range of potential answers to this question and therefore a wide range of activities that your consortium could carry out in order to make the added value generated by your project more attractive to potential funders. The four main addition-ality arguments are market failure, knowledge spill-over, know-ledge transfer and criticality of funding.

Market failure

In an ideal world, there would be an economic demand for your service or product to be improved. This would lead to a desire for you and your partners in the market to invest in developing new innovations that would satisfy this demand and provide you with a healthy return on your investment. Often this is the case but not always. Market failure is said to have occurred when you and your partners in the market cannot invest sufficiently in the development of new innovations and the pace of improvement slows beyond what is optimal for the economy.

Perhaps the greatest contributor to market failure is that research, development and innovation projects are characterised by a high degree of risk and uncertainly. Companies have limited resources to spend on research and development and often do not have the confidence to direct those resources at projects that may not be successful. By helping to fund the project, the funding body is sharing the risk. Ideally the level of funding will be just sufficient to entice the companies to carry out the project, any higher is wasteful of the grant resources, any lower and the project would still not go ahead.

Knowledge spill-over

Research, development and innovation projects often generate benefits for society in the form of knowledge spill-overs that are not of any benefit to the project participants. Left to the market, the project might not have an attractive rate of return for the participants alone. The benefits to the wider society is only achievable therefore, if the funding body makes part of the investment, thus increasing the rate of return for the participants. To achieve this knowledge spill-over however, the project must disseminate the results of the project so that society can indeed benefit.

A logical extension of this argument comes in the creation of general knowledge through fundamental research. General knowledge is created when research leads to results that cannot be protected or exclusively exploited by the participants. The likely return on investment for unprotected research is often very small for the participants and so would be unlikely to occur unless state funding was provided. This is why states often have to pay 100% for fundamental and pure research. Here again the dissemination of the results to a wide audience will be an important part of the project.

Knowledge transfer

In many cases, the innovation that is required in the market is developed within a university or similar research organisation. It is very seldom the case that companies can simply buy these innovations off the shelf as they usually require further development before they can be applied to a real product or service. In these situations, the company needs to work with the university on a project to develop and transfer the knowledge to the company. This process works best if the academic experts are fully involved in the project but they are seldom happy to leave their academic career to join the company.

The answer is to run a collaborative project between the company and the university but this is expensive. The university, quite rightly, will need to see all its costs and overheads covered for the work it will have to carry out. Universities often want to involve post graduate and post doctoral students to do the necessary work and to add to their research capability, all of which puts the cost up without necessarily improving the companies return. A funding body is then required to share the costs to an extent that makes it favourable for the company to go ahead with the project.

In reality, most knowledge transfer projects are the result of the state wanting to make a return on its investment in fundamental research by providing knowledge transfer grants that promote projects to develop and commercially exploit earlier research work.

Also under this heading comes the justification for knowledge networks and coordination activities. These special projects often help to educate the commercial players about the technology that could be available to them, help coordinate research activities across a market sector and to help companies and universities find partners with whom they can work on innovation projects. It may help your grant application if you identify relevant knowledge networks that you will use to disseminate your results.

Criticality of funding

It is not always a good idea to tell a funding body that without their support the project will not go ahead in any shape or form. This gives the impression that you are more interested in their money than the project's results. Ask yourselves honestly what could be developed even if there was no external funding available.

The criticality of funding argument states what difference the funding will make to the proposed project and the impact that will have both commercially and for society. There are many potential impacts such as:

- Time to market – without the additional funding the project will move at a much slower pace, potentially missing the window of opportunity in the market or allowing outside competition to take hold.
- Scope of the project – without the additional funding the project will not be able to address all the key issues, which would lead to a sub-standard solution.
- Higher risk – without the additional funding the project will be less innovative and not take full advantage of what could be possible.
- Smaller team – without the additional funding the SME or university partners will not be able to afford to take part, so that their knowledge will not be available and they will not benefit from the innovations developed.

Sometimes it will simply not be possible to progress the project at all without additional funding. If this is the case make sure you carefully explain why this is the case in your proposal.

Technology readiness levels

Technology readiness levels are used to describe the position of a piece of technology on a scale between pure research and eventual successful deployment. This scale was originally developed by NASA in the late 1980s and has subsequently been adopted by many government departments around the world (especially defence departments) to describe the status and risks involved in research and hence at what rates it should be supported.

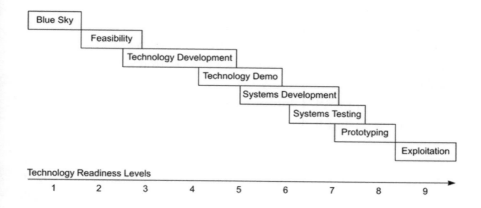

Fig. 1.2. Technology readiness levels (Source UK MOD)

The scale is illustrated in figure 1.2 and uses numbers from 1 to 9 to describe the current status of a technology and the types of research work involved at each stage.

1.3 Eligible costs

The vast majority of collaborative research grants provide a percentage contribution towards the costs of the project rather than providing a lump-sum and letting you get on with the work. Therefore in order to know how much to ask for, you need to know how much it will cost to carry out the project and what costs the funding body will accept as eligible. Again it is impossible to cover every rule from every scheme here, but what follows are the general guidelines that are common among funding bodies. You must check the guidance for the grant you apply for in order to understand their specific criteria.

A fundamental rule that is almost universally applied is that all the project costs should be real and not include any profit elements. In other words only the actual costs accrued by the company in carrying out the project are acceptable. For example the cost of a days work by an engineer is not what they would be charged out at to a client, but what it actually costs in salary and package costs to have them work for that day. A lot of mistakes are made when this rule is applied to products, services and in particular software licences. If a company supplies a software licence to a partner in the project to work on the project, the eligible cost is not the normal market price of the software but what it actually cost to provide it. This might include burning the media and postage but not much else!

Labour costs

The largest costs in research and development projects are usually the labour or human resource costs. These are the salaries of the people working directly on the project. Typically it is appropriate to calculate these costs on a person-day basis. You can usually include direct employment costs in this calculation such as pension contributions and employment taxes but not any profit distribution, share options or dividends that they receive from the company.

To work out the person-day cost, take the total annual salary cost for the individual and divide it by the number of available working days in the year. To do this, subtract weekends, public holidays and annual leave entitlement from 365. An example is shown in table 1.1.

	Direct Labour cost	Available working days	Eligible labour costs
Salary	50,000		
Pension @ 9%	4,500		
Employment Tax @ 11%	5,500		
Total annual direct costs	60,000		
Days in the year		365	
Weekends		104	
Public holidays		8	
Annual leave entitlement		20	
Total available working days		233	
Day Rate			258
Days to be worked on the project			400
Total Eligible Labour Costs			103,200

Table 1.1. Calculating direct labour costs

To calculate the cost of part time staff, first bring their salary level up to a full time equivalent. For example if you pay someone

$10,000 a year but they work one day a week or 20%. Their equivalent full time annual salary would by $50,000.

Overheads

In addition to the direct labour costs, most funding bodies allow a contribution towards the overheads of the company. Overheads are the costs associated with employing someone that are not included in the direct costs. Overheads might therefore include the desk they sit at, the energy they use and so on.

Because it would be unrealistic to try and calculate these costs on an employee by employee basis, overheads are usually calculated for the whole company as a percentage of the whole salary bill then applied to the project staff costs on that basis. For example if a company has a total salary bill of $1,000,000 and the overhead costs total $800,000 then the overhead is 80%. That means that for every $1 the company pays in salary it spends an additional 80c on overheads.

The eligible overheads for the example in table 1.1 would be 80% of $103,200 or $82,800.

This is fine in theory but what should be included in the overhead calculation? Here, once again you will need to check with the individual funding bodies but the following items are usually acceptable:

- Heating, lighting and general utilities
- Support staff salaries such as administration, human resources and senior management
- Staff welfare and occupational health programmes
- Staff recruitment, training and induction costs
- General office equipment, IT services
- General laboratory costs and stock
- Corporate legal and accounting fees
- General postage and communication costs

- Building maintenance, rent and security costs

The following items are generally not allowed as part of the overhead calculation:
- Product of corporate marketing costs
- Entertainment costs
- Anything connected with product manufacture, delivery or client services provision
- Bonus schemes or profit distributions
- Company cars

To be practical, most overhead calculations are provided based on the previous year's accounts of the company. In this way real numbers can be used. Once a rate is arrived at and agreed by the funding body, it tends to remain fixed for the duration of the project.

Other eligible cost headings

Other common eligible cost headings allow for the purchase of materials that will be used during the project, capital equipment that might need to be purchased, travel and subsistence costs associated with the project and any sub-contracted work that is required.

When detailing with materials costs take care not to include items that you have already included in the overhead calculation such as stock chemicals in your laboratory, stationary and so on. A common mistake is to include office computers in the overhead calculation and then claim the purchase of a PC for the engineer working on the project. Either exclude IT costs in the overhead calculation but add the PC to the equipment costs or vice versa.

Capital equipment costs are normally eligible but may be restricted to their depreciated value and utility. The depreciated value is the difference between what it cost to purchase at the beginning of the project and its residual value at the end of the project. The equipment's utility may also be part of the calculation. If the equipment is only used on the project, then the entire depreciated value is eligible, if however it is used 50% for this project and 50% for a different one, the eligible cost to the project will be 50% of the depreciation value.

There are a few exceptions to this equipment cost rule. In the UK, academic purchase of equipment is now treated as the purchase price with no utility or depreciation reduction. The reason for this is that the UK research councils use grants to update and improve the equipment base within the universities. Before you immediately decide to put all equipment purchases on the university's costs however, remember that the equipment must remain on the university campus and remains the property of the university when the project is completed.

Many funding bodies accept sub-contracted work as eligible costs. Sub-contractors could carry out some non-innovative aspects of a project such as fabricating circuit boards or carrying out specialist testing. They would be paid full price for their contribution but not receive any benefit from the project outcomes.

It may be possible to include other costs such as patent filing to protect the intellectual property you generate, commissioning market analysis or purchasing specialist training for the development teams. If you don't ask you won't get.

Most grants, once awarded, are fixed and so if you forget to add something to the proposal; it won't be eligible during the project. It is therefore very important to estimate your costs as accurately as possible. We will visit this again later in the section on project planning.

1.4 Financial management

Once you have worked out how much the project will cost, it is important to understand how the grant will be paid out. It would be nice if the funding body were to send you a cheque for their contribution at the start of the project and leave you to it, but this is not how they tend to work. The majority of funding bodies will pay their contribution in instalments and often in arrears. This means that at the end each period (usually an annual quarter), the funding body will expect a statement from you, detailing what you have spent on the project. They will then send you funds equal to their contribution of the costs. So for example if your grant is a 50% contribution and you tell them you have spent $1,000 they will send you $500.

This puts pressure on your company cash flow because you must pay the full amount for the research up front each period before you get paid back by the funding body. Remember too that the payment process may not be immediate; it may take you a little while to calculate and present your costs after the period has ended. The funding body will then take some time to check these and process the payment.

Further delays may be experienced if the grant is paid to a lead partner who then distributes it among the other partners in the project. Make sure your finance department appreciates this from the outset.

Keeping records

It should be obvious, but if you are required to make statements about what you have spent on the project, you need to keep good records of your expenditure and activities as they occur. With labour normally being the largest expense, timesheets may become necessary to log activity on the project. Most large companies will have systems to log employee time, but make sure any SME partners have sufficient measures in place to be able to track their expenses.

Academic partners can find difficulty with concepts like timesheets. Make sure that they are able to comply with the accounting terms of the grant.

Audit requirements

To avoid fraud, most funding bodies require an audit of the project expenditure. This normally takes the form of an independent auditor examining the records of the project and the accounts of the partners to confirm that the costs claimed have indeed been properly spent. The details of these audits vary considerably, some funding bodies will require an independent audit with every claim, and others will just want one at the end of the project.

These audits can be expensive, so make sure you understand the exact requirements and shop around for the best deal. In theory your existing company auditor should be cheapest because they already understand your accounting procedures but this is not always the case. In some cases it might make sense to have one auditor carry out the audit for all the partners rather than have each partners' costs audited separately.

Monitoring your spending

As already explained, most funding bodies will provide a grant in the form of a percentage contribution towards the total cost of the project. It is then normally up to the partners to decide how this is shared out (there are however likely to be rules about the maximum each partner is allowed to receive). For example most universities will require 100% of their costs to be met by the grant as they have no other way of funding the work.

In extreme cases, all the grant money could be going to universities. For example, imagine a project with a total cost of $500,000. The project is made up of a university and a company, each with costs of $250,000. If the funding body provides a 50% grant then the entire grant will go the university. However, there will only be enough grant available to cover the university's costs if the company does indeed incur costs of $250,000. If they under-spend and only incur costs of $200,000 the total project cost will be just $450,000. In this case the 50% grant will be reduced to $225,000 leaving the university short.

It is therefore very important when estimating the costs up front, to be realistic and ensure that everyone will be able to accumulate the costs they initially commit to the project. Once running, it is also very important to monitor how the project spend is progressing. If some partners are under-spending it may be possible to move more work to them from another partner or allow another partner to carry out more work to take up the slack.

1.5 Grant contracts and offers

When you enter into a grant funded project you enter into a legal arrangement both between you and your partners but also between you and the funding body. In some cases this legal arrangement will be set out in a contract that both you and they sign. On other occasions the terms of the grant will be set out in an offer letter, which by signing you indicate your acceptance of its terms.

There are some legal differences between contracts and grant offers. Strictly speaking a contract ties you into delivering the project and its results in return for the grant. An offer letter is subtly different in that you are not contracted to do the work, but if you carry out the project within the terms, a grant will be provided. The differences are not worth worrying about too much however, because the choice will not be up to you, it will be imposed by the funding body.

There are also differences in the structure of contracts or offers made. In some cases a single contract is made with the lead partner of the project. The other partners will sign their acceptance but the grant process will be run through this lead partner. When claims are made, it will be made through the lead partner who will receive the grant and be responsible for distributing it to all the other partners. This is known as a 'hub and spoke' contract and is quite common. Alternatively the funding body could provide each partner with an individual contract and allow each to make grant claims independently. These 'individual' contracts are often used if a project is phased and partners' contributions occur at different times throughout the project. Again the choice of structure is likely to be imposed by the funding body.

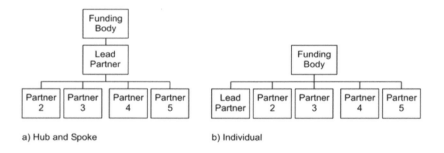

Fig. 1.3. Contract structure options

Figure 1.3 illustrates these two grant contract structures.

Grant terms and conditions

Whether your grant is provided through a contract or an offer letter, the terms are likely to be broadly similar. There will be an active period for the project between a start date and an end date.

Normally only project costs incurred between these dates will be eligible for a grant contribution. In other words, if you start work before the formal start date, you won't get paid for it. The terms will also set out in detail the grant payment process and the requirements for any auditing of the project costs.

Being a legal document, the terms will include a set of measures to control what happens when everything goes wrong. There are likely to be terms to cease the grant if you or one of your partners become insolvent, fail to carry out the project to their satisfaction or wish to terminate the project. There are also likely to be terms that cover the creation of intellectual property and confidentiality.

There will also be terms that ensure that the funding body's objectives for providing the grant are met. In the case of a state aid grant to promote the creation of new commercial opportunities, there are likely to be terms that compel the partners to use their best endeavours to exploit the technology within a certain time period. There may also be terms that restrict the exploitation to certain territories. For example, if the grant is given in an attempt to create employment in a depressed area, then any jobs created as a result of the project will have to be in that area.

Many grants also include terms that compel the partners to disseminate their results in the form of publications, or take part in networking activities with other projects funded by the same body.

Finally the terms will include any arrangements for you to report the progress of the project and for the funding body to monitor the project.

The grant terms often work in concert with a Consortium Agreement or a framework document which sets the terms of your collaboration with your partners. We will look into these in a lot more detail in the next chapter, but for now be assured that you will need one of these documents. Normally the text will be for you to agree with your partners but some schemes provide model agreements to get you started, others will impose an agreement structure on you.

> **Read the terms first!**
>
> The grant terms might not be the most exciting read you will ever have, but it can save you a lot of effort and embarrassment if you can read them before you submit your proposal.
>
> There is nothing worse than winning a grant only to find that your consortium then falls apart because one of the partners cannot accept the terms. All that effort in writing the proposal is wasted and it does not do your reputation with the funding body any good at all.
>
> Always ask the funding body for a copy of the terms up front and make sure all the partners have read them, shown them to their management, and are happy to sign an offer letter if you are successful.

Reporting and monitoring

Depending on the size of the grant, the funding body will impose some form of reporting structure on you and your partners. This could be as simple as a short report when you have finished, or it could entail detailed reporting coupled with a visit from an appointed monitoring officer every month.

It is worth finding out in advance what level of reporting will be required. If it is particularly arduous you will need to include an estimate of the time it will take in the labour costs of your project. By way of example, table 1.2 illustrates a three year project comprising a collaboration of five organisations, which is monitored through reports and a project management meeting held every quarter, the total management effort is likely to be almost 6 person-months! If we use the day and overhead rates from table 1.1, this equates to over $50,000 of costs.

	Days effort	Total days effort
Report from each partner each quarter	0.5	
5 partners for 12 quarters		30
Partners attending quarterly progress meetings	1	
5 partners for 12 quarters		60
Project manager chasing and compiling reports each quarter	2	
12 quarters		24
Grant total		114 days

Table 1.2. Reporting efforts for a typical project

It is important to remember however that this effort is critically important for the success of the project. It is time well spent for each of the partners to know what the others are doing and to be able to discuss progress on the project. In other words, you should be doing this anyway, the imposed reporting merely formalises the process.

1.6 Winning grants

What follows in this section is a set of best practice advice and tips for increasing your chances of success in winning grant support for your project.

Pick the right grant

There is actually a great deal of choice when it comes to grants. They are available at local authority, national and international levels from the public sector. They are available to build collaborations between countries. They are available to support regional development. They are available from charities to support social, environmental and medical innovations and they are available for defence projects. They are even available from wealthy philanthropists in areas that excite them.

Each funding body will have a clear idea of the sort of project they want to support to meet their particular objectives. The first thing you need to do is make sure that your project fits these needs. There is absolutely no point putting in a proposal that does not meet their criteria 'just in case'.

The other side of this coin is to make sure that the grant suits your needs too. Making sure that you can live with the grant terms is part of this process but if, for example, the grant is dependent on you working with a company from a different country, make sure that this will add value to your project before going ahead.

The level of pre-competitiveness of your project also needs to match what the funding body is trying to achieve. It is in everyone's interest that this match is made, if you are forced to move your project away from the market to suit the grant, you could end up no further forward than you are now.

Instructions and briefings

Most grant schemes provide documentation with instructions and guidance for applicants. You and your partners should read this information. It sounds obvious, but many applicants do not bother to read this guidance and then wonder why their proposals are criticised and rejected. You should study all the guidance very carefully and refer to it while you are writing your proposal. This is particularly important when the assessment is based on a defined set of questions rather than a free flowing proposal document.

As freedom of information becomes more commonplace, it should be possible to obtain any guidance that the funding body provides to their assessors. If you can, read this guidance in conjunction to the applicants' guidance to ensure you understand fully what the assessors will be looking for.

Some larger schemes provide briefing sessions for applicants. You should definitely make every effort to attend these sessions. You are likely to pick up insights into the objectives of the scheme and what the assessors are looking for in the proposals. They also give you the opportunity to ask direct questions and seek individual advice.

If possible bring as many of your partners along as you can and take the opportunity to discuss the project and the details of your proposal. That way when you come across unclear areas you can get them resolved at the briefing.

> **Get some face time**
>
> If the scheme offers a briefing session than make sure you talk to the key officials from the scheme. Don't pester them, but find a way of introducing your project to them. Very often you will get advice on the shape of your project and candid suggestions about how it could be improved to suit the scheme. After all it is in the official's interest to attract good proposals; they lead to good projects and successful outcomes for the funding body.
>
> If there are no formal briefing sessions, and for smaller schemes this is often the case, the officials are often available for private meetings to help advise potential applicants. If

possible arrange a meeting to discuss your project idea and how it might fit with their scheme.

Get the basics right

The funding body will have outlined exactly what they expect to see in your proposal but there are a small core set of areas that they are all likely to require. These are:

- why you need their funds,
- what innovations are you going to make,
- what are you going to do,
- what will happen to the results,
- and how will you manage progress.

The first point is the additionality question already discussed. In describing the innovations you are going to make, you will need to ensure that what you are proposing is indeed novel. If there are similar technologies or methods either proven or under development elsewhere, you will need to explain very carefully how your ideas are different and worthy of additional attention.

The innovations may be required to fit within a technical scope that the funding body is trying to encourage. If this is the case make sure that you describe the innovations in this context and stress the relevance of the innovations to the academic subject or industry sector that the scheme is attempting to support.

Describe what you are actually going to do during the project. This is often referred to as the technical approach or method. This area of the proposal often separates the well thought out projects from the 'punts'. It is not uncommon for assessors to report that a proposal sounds very exciting but that they don't actually know what the partners are going to do. We will cover project planning in more detail later, but for now ensure that what you write allows the assessor to appreciate what the deliverables of the project are going to be and exactly how you are going achieve them.

The dissemination and exploitation aspects of your project will be very important to most funding bodies because this is how they achieve their knowledge spill-over and economic growth objectives. The basic difference between dissemination and exploitation is that you receive a direct return for exploitation of the results.

Dissemination often includes the traditional academic journal and conference publications but do not forget the commercial and industrial audiences. These people don't read academic journals or attend conferences, but they do read the industry press, attend professional institution seminars and participate in networks. Funding bodies often support specific knowledge networks that disseminate results from the projects they fund. You may be compelled to disseminate through such networks, but even if you are not, it is a good idea to include them. It will please the funding body and give you and your partners' exposure that will enhance your image in the market.

Exploitation plans at this stage will be limited to a description of how you expect to take the results to market at the end of the project. Even if it is not explicitly requested it is a good idea to outline the size of the potential market opportunity for the innovations and describe the current and expected dynamics of the target markets.

Project management is an essential part of any project. Table 1.2 has already illustrated the effort required simply to report your progress. To actively manage the progress takes at least twice this effort. The funding body will want to assure itself that it is investing its funds in safe hands. We will cover project management later but for now ensure the project management methods are explained clearly and that items such as Gantt charts are clear, demonstrating that you know what you are doing.

Build the right team

You could have an idea for a project which promises outstanding innovations, has a vibrant waiting market, will have sound management methods and compelling additionality arguments, but if the team you build is not right you are still going to fail.

Pulling different companies and organisations together into a team with compatible objectives is not easy at the best of times. Trying to do it as a proposal deadline looms, expecting them to trust each other with their plans, work together for several years and after all that, cooperate as they exploit the results is even harder.

If at all possible build your collaboration in plenty of time. This will give everyone a chance to get to know and trust each other at a sensible pace before the proposal is submitted. Networking events and conferences are good places to meet prospective partners but sometimes cold calling is necessary. If it comes down to this, try to get into a prospective partner through a contact of some type. A call to a contact to ask who you should talk to in their company about the project and whether they could help with an introduction will probably be much more fruitful than simply emailing the Chief Technology Officer.

Partnering activities are arranged by some funding bodies and industrial networks. These can be fruitful if you really are stuck for partners but make sure you properly research prospective partners before contacting them. There is a risk that they have added themselves to the list to find out what the competition is doing rather than through a genuine desire to collaborate.

The team will grow through this process and the project will continue to redefine itself as each new set of objectives and capabilities joins the table. Be prepared to let this happen but ensure that each partner remains happy with where the project is going and that their objectives are still likely to be met. As soon as possible start working on the project plan including the deliverables and the major technical and market hurdles. Compare the capabilities of the team with these plans; any holes will need to be filled with additional partners. When the team is complete you can start planning in more detail.

Not all members of the team need to be full partners in the project. It is not uncommon for funding bodies to accept sub-contractors to fulfil non-core aspects of the project. Using sub-contractors means you can keep your core team small and manageable but fill any gaps in your capabilities. Using sub-contractors can be expensive of course, but since they do not receive any share of the project benefits or the intellectual property they may work out cheaper than an additional partner in the long run.

End users are often an essential component of a research and development project. They ensure that the project deliverables are aligned to a clear commercial and practical need. Many projects, particularly far from market projects, have many potential end user groups. It could be unwieldy to include too many different end-users in the core project team, but fortunately there is a solution to this dilemma. Advisory groups are an excellent way of ensuring market focus for the project without making the team too large. An advisory group is ideally made up of between six and ten people who come together every six months or so to hear about progress with the project, to see and comment on demonstrations and to provide feedback and market intelligence to the project team. The group can comprise a mixture of industrial and academic members including the key opinion leaders in the field.

Forming a good advisory board is actually easier than it sounds as the time commitment and hence your costs are very low. The group members benefit through keeping up to date with the exciting things you are doing and at the end of the day it is rather flattering to be asked.

Get the team together

Once you have your collaboration organised make sure you all meet together at least once while the proposal is being written. The meeting will be very important for planning the project and ensuring everyone agrees with it. It will also give everyone an opportunity to get to know each other a little so that if any problems occur, you have a chance to fix the issue or find a replacement.

Perhaps of most importance though is that it allows everyone to show their commitment to the project. Taking a day out to travel to a meeting is a trivial undertaking compared to the effort that will be required during the project itself. If you can't get the team together to work on the proposal, it does not bode well for the future, does it?

Interviews

Many funding bodies employ interviews or visits as part of their assessment process. If this is the case, be under no illusions that the interview is a critical part of the process and will determine your ultimate success. In other words, the written proposal merely gets you onto the shortlist, your performance at the interview is the real deciding factor.

Given this level of importance it is imperative that you make a good showing. There are two key elements that will help:
- bring the right people
- provide a good presentation

You may be limited to a maximum number of people you can bring to the interview. If not make sure all the partners are represented. If you are limited, think carefully about the team you bring. Think about the objectives of the funding body and tailor your team appropriately. If the funding body is keen on exploitation, don't fill the room with academics at the expense of the end users in your consortium.

It is also a good idea to bring the individual who will be responsible for managing the project. Let them explain how they will run the project so that the panel can see their money will be in safe hands.

A good presentation is obviously essential, it is a good idea to have two people doing the talking, but bring the others in during the questioning session so that everybody has a chance to speak. This is likely to work much better than trying to involve everyone in the actual presentation which will end up being bitty and lose its flow. Practice the presentation and try to anticipate what questions are likely to be asked. If you are given a time limit for the presentation, make sure that you stay within it. You might think that what you have to say is worth running over time for, but the panel won't.

Remember also that you are talking about a proposal level plan. If the panel makes suggestions or spots errors, don't stand defiant, but rather take them on board, thank them for improving the plan and let them know that you are receptive to advice. After all if you included an advisory group, you need to show that you are capable of listening to what people have to say.

Finally, your presentation is not just about the project it is about you and your team. It may be old fashioned but do dress smartly. A professional appearance shows respect to the panel and sets a level of expectation when you enter the room. First impressions could be very important.

Learn from disappointments

The cruel truth is that there are many more good projects than there are grants available. Most large schemes are only able to fund one in every ten or more proposals they receive. This means that if your proposal is rejected it may not be because it was bad. It could have been very good but unfortunately there were too many others that were better.

Most schemes provide feedback in the event that you are unsuccessful. It is not easy but try to take the positives from the feedback and work on the negatives. If possible try to meet the officials to discuss how you could improve next time. It is very rare that there are any appeal procedures as in reality the available funds will all be committed to the winners. By the way, it is never a good idea to show your anger to the funding body or start an argument with them.

Think back to your additionality arguments and what you might be able to get started even without their funding. If the project really is so important to you and your partners, make a start where you can while you keep looking for the additional funding.

1.7 So in conclusion

1. Grants can be won from:
   - Governments, (local, national and international)
   - Charities
   - Philanthropists

2. State aid regulations define what grants can be awarded by governments within Europe including:
   - Special support for SMEs
   - Levels of funding for different levels of pre-competitiveness
   - Support for regional development

3. Additionality is a key consideration and includes:
   - Market failure
   - Knowledge spill-over
   - Knowledge transfer
   - Criticality of funding
   - Technology readiness

4. Typically project eligible costs will include:
- Direct labour costs
- Overheads
- Materials
- Capital equipment depreciation
- Travel and subsistence
- Sub-contracted work

5. Good financial control is required when running grant funding projects including:
- Keeping financial records
- Ensuring that you can satisfy any audit requirements
- Monitoring and controlling spending

6. Grant contracts and offer letters set the legal framework under which you can receive the grant.

7. Steps towards successfully winning grant support include:
- Picking the right grant to apply for
- Following the instructions
- Attending the briefings
- Getting the basics right
- Building a strong partnership
- Learning from disappointments

2 Legal arrangements

Whether you plan to run a grant funded project or not, all collaborative R&D projects require some form of legally binding agreement between the partners that controls their relationship both during and after the project. This document goes under various titles, they are commonly referred to as consortium agreements, collaboration agreements or framework for collaboration documents. Regardless of the name, they all do basically the same job.

Negotiating and agreeing a consortium agreement can be complex and time consuming. You are therefore strongly advised to start discussions as early as possible in the process. It is not unknown for it to take over a year to agree a consortium agreement. Sometimes grant winning projects fall by the wayside because the partners cannot reach agreement in time to accept the funding offer. If you can at least agree on the basic principles or reach a 'heads of agreement' before the grant proposal is submitted, this will greatly ease the pressure.

Obviously these agreements are legal documents and as such you will need to seek professional legal advice and assistance in preparing and negotiating the terms.

I appreciate that this is not one of the most exciting or sexy topics but it is one of the most important aspects that you need to deal with before you can get start creating and running your stimulating, innovative and potentially highly beneficial project. So, get some coffee, a chocolate biscuit or two and let's get going.

2.1 Purpose of the agreement

The consortium agreement sets out the internal management guidelines for the consortium and the project. Even if your project is grant funded, in most cases the terms of the agreement will be for the consortium to decide. That said, many funding bodies offer model agreements that could greatly speed up the process by providing a good starting point.

At a basic level, the consortium agreement should include terms that set out:

- the definition and scope of the project including a description of each partners' contribution,
- how the project will be managed including the roles of steering committees and the project manager,
- what responsibilities and liabilities are imposed on each partner,
- how additional partners are allowed to join the consortium,
- how partners may withdraw from the project,
- how the budget for the project is managed and how any grant funding is distributed,
- how background and foreground intellectual property is dealt with,
- how partners are allowed to disseminate the results of the project,
- how disputes are resolved,
- how the project can be terminated and
- what happens when things go wrong, including legal 'boiler plate' such as force majeure, disclaimers and so on.

The following sections describe each area in more detail.

2.2 Agreement terms

Scope of the project

This part of the agreement describes the scope of the project, its objectives and the approach that will be taken to achieve them. This is commonly provided by referring to the project plan which would be attached to the agreement in the form of an annex or schedule.

This is a major part of the agreement because it basically sets out what each partner promises to do during the project. It also sets out the resources that each partner will commit to the project. It would normally also refer to a project budget setting out what each partner will spend in terms of cost headings such as labour, materials, equipment and so on, as well as any cash contribution that is being paid to any of the other partners. We will cover what should be included in the project plan in much more detail in the next chapter.

If the project is grant funded, a copy of the proposal and the grant offer letter or contract should also be included in the schedules.

The scope description should also describe the legal hierarchy of the various project documents. Any grant offer terms will usually take priority over the consortium agreement. This means that if a term in a grant offer letter contradicts a term in the consortium agreement, it is the term within the offer letter that shall apply. It is therefore simpler to make the consortium agreement agree with any offer letter terms to avoid potential confusion (and legal bills) in the future.

Responsibilities and liabilities

There should be a set of fairly standard clauses that compel each partner to carry out its respective contributions to the project as defined in the project plan, act in good faith and comply with their obligations as defined in the rest of the agreement. If the project is grant funded, you may also want to add terms that ensure that each partner complies with the reporting and monitoring requirements set out by the funding body.

The various liabilities may take a little more effort to agree on. You are likely to need terms that outline and restrict each partner's liabilities to the others in dealing with background and foreground intellectual property, the supply of materials and so on. For example you may want to include a term that prohibits a partner from supplying background IP to the project in the knowledge that by using it within the project the rights of a third party would be infringed.

Limits of any indemnities and warrantees between the partners will also need to be set. These liabilities may result from acts or omissions by each partner that may cause damage or loss to the others.

Liabilities to third parties will also need to be covered by the agreement. If the liability to third parties is joint and several, injured parties may sue any consortium member for damages, whoever was actually at fault. This could lead to injured parties simply attempting to sue the richest partner rather than the partner at fault. Not surprisingly the lawyers will argue strongly about such terms and hopefully arrive at a set of wording acceptable to everyone.

Project management

The most common way of managing a collaborative project is through the creation of a project steering group or committee. Normally each partner will nominate a single representative to this committee but may be entitled to have others attend meetings as non-voting observers. Each full member of the committee has a vote and all significant decisions are made by a simple majority. This is a very democratic approach and does not take into account the relative contributions of each of the partners, so in extreme cases you may need to allow a major partner to have two or more votes. The terms should define this steering committee including its voting rights, how regularly it meets, what constitutes a quorum, the appointment and powers of a chairperson and so on.

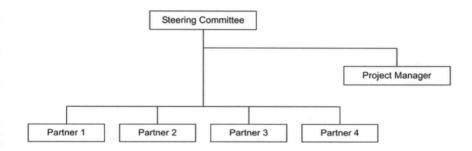

Fig. 2.1. Typical project management structure

It is also usual that the steering committee will appoint a project manager who will be responsible to the committee for the day to day running of the project, liaison with the funding body, monitoring and reporting the progress of the project with respect to milestones, deliverables and resources. The terms should describe this role together with their appointment and replacement mechanisms.

Changing the consortium

It is not unusual for the consortium to change during the life of the project. Company strategies and priorities change, academic staff move institutions taking research groups and projects with them and so on. The consortium agreement therefore needs to make provision for partners to leave the consortium and for new members to join.

The addition of new partners is the simplest to control. Usually the addition will need to be agreed unanimously by all the existing partners and once in, they have to agree to be bound by the terms of the existing consortium agreement. In some cases, new partners may be required to make a contribution on entry which recognises the resources already expended by the existing partners. The exact nature of the contribution or value of any sum to be paid would be agreed by the steering committee.

The terms relating to the withdrawal of partners from the project and agreement are likely to be more arduous as ownership and access to new intellectual property and the disruption caused to the remaining members can be difficult to deal with.

Typically, partners can only leave the project with the unanimous consent of the other members and subject to any conditions that they might unanimously agree to impose. These conditions might include a sum payable by the departing member as compensation to the others, or an agreement concerning the continued use of any intellectual property that has been generated or was provided by the departing partner.

In the event of a partner withdrawing from the consortium, the agreement will need to allow the steering committee to reallocate their obligations and contributions towards the project, both within the remaining consortium and through the introduction of new partners or sub-contractors.

Termination

Termination clauses usually deal with two eventualities, the removal of a partner who is in breach of the terms of the agreement and the winding up of the whole project.

The removal of a defaulting partner will normally involve the non-defaulting partners unanimously instructing the chairman of the steering committee to issue written notice of the breech, stating a period in which the breech must be remedied to avoid the removal of that partner from the project. Similar clauses will deal with the winding-up or insolvency of a partner.

In the event that it is agreed unanimously that there are no longer valid reasons to continue the project, clauses should allow the termination of the project. If the project is grant funded, the terms of the offer letter or grant contract will be relevant in this situation and so should be considered. It is also worth noting that some partners may not wish to terminate a failing project for their own reasons. Academic partners for example who have employed research staff and are receiving 100% grant funding are unlikely to agree to a termination that would leave them with outstanding liabilities. It may therefore be necessary for the unanimous vote of commercial partners to be sufficient to wind the project up.

Financial management & grant distribution

The steering committee should oversee the financial management of the project. This is likely to include regular statements of expenditure and resource commitment from each member set against what was initially agreed in the project plan and budget.

Grant distribution terms will be required for projects supported by a funding organisation. The terms should set out time periods for financial disclosures and grant requests. In the case where a lead partner is claiming grant on behalf of the consortium and then distributing the grant on receipt, terms will be required which set out the invoicing regime and the time periods in which the lead partner has to make grant claims and distribute the grant when it is received.

Terms may also be required that permit an independent accountant or auditor to examine books and records of the lead partner and project manager on demand.

Intellectual property rights

The majority of R&D projects create new innovations and hence intellectual property that can be owned, protected and traded. The collaboration agreement will need to make provision for the ownership of this new intellectual property and for existing or background IP that is supplied to the project. In the case of US state funded projects where the IP is disseminated to the wider community, the agreement may need to make provision for the identification and dissemination of the new intellectual property.

What are background and foreground IP?

Background intellectual property is IP that was generated either prior to the start of the project or developed independently of the project. Background IP is made available to the project either for the project's development or as a necessary part of the exploitation of the project's own IP.

Foreground intellectual property is the IP that is generated as part of the project's R&D activities.

It is good practice to maintain a log of intellectual property that is used and generated by the project. This log should be maintained by the project manager and be accessible by all partners.

Normally the ownership of new intellectual property vests with the partner that created it and they are then responsible for its protection. If a partner is not able or willing to protect its foreground intellectual property, the steering committee may be given power to consider the assignment of rights to a different partner who has the desire and resource to protect them. In many cases, new IP is generated jointly and in such cases the ownership would also be held jointly. The owners then agree between themselves how the rights are to be protected and paid for.

The ownership of intellectual property is only half of the issue. The use of it within and after the project is just as, if not more important. Typically background IPR needs to be assigned on a non-exclusive, royalty-free base by each partner to the other partners, to enable them to undertake the project. New foreground IPR needs to be equally assigned for research purposes but not for commercial exploitation purposes. If one partner requires use of another partner's foreground IPR for commercial purposes, the terms of the agreement need to ensure that this can take place. This will typically be through the payment of a license fee. The level of these fees could be agreed in advance, but are more commonly referred to as a fair and reasonable rate taking into consideration the respective financial and technical contributions of the partners in developing the IP and the costs incurred in its protection. This is a form of 'agreeing to agree' and defers the more difficult negotiations into the future when more will be understood about the ownership and worth of the innovations that are developed during the project.

The generation of new technology or innovations and hence intellectual property is the main reason why companies enter into collaborative R&D projects and so this area of the agreement is often considered the most crucial. It is therefore an area around which most of the negotiations revolve and where most of the lawyers' involvement is concentrated.

Publication and announcements

If you have academic partners within your consortium, they are likely to be very keen to publish journal and conference papers on the project and the results. Many companies also benefit from issuing press releases and publishing articles in the trade press about the exciting and leading edge projects they are involved in. Small companies will also been keen to tell people that they are working with large and prestigious companies as a way of impressing their clients and investors.

Tight control over publications and announcements is required where they could harm the potential protection of new intellectual property. For example, patenting becomes very difficult, if not impossible, once the technical details have been published. The terms of the consortium agreement should therefore include control over publications. Typically, this is done by requiring unanimous agreement on the text of intended publications prior to their release, together with clauses that state that permission should not be unfairly or unreasonably withheld. Time limits might be set to allow sufficient time for partners to read and agree a text prior to an intended publication date.

If academic research students are involved in the project, there should be provision that nothing in the agreement shall prevent them submitting a thesis to examiners in order to graduate.

Confidentiality

One of the dangers of entering into a collaborative project is that you have to share sensitive and confidential information with your partners. This may not be limited to technical know-how but may include market intelligence and future business plans. Each partner needs to be assured that their secrets will only be used by the other partners for the purposes of the project and shall not be disclosed outside of the consortium. The terms of the agreement will need to compel each partner to treat all received confidential information with the same care as they treat their own information and to take reasonable steps to ensure its continued confidentiality.

It is also quite common to put a time limit on the duration of the confidentiality clauses. For example it may state that the information shall remain confidential for a period of 10 years from the time of disclosure. In most cases there will also be let out clauses should the information become public through another means, be acquired lawfully from another source or is required to be disclosed by law or by a court order.

Clauses may also deal with breaches in confidentiality which could, for example lead to the guilty party being expelled from the project.

Boiler plate

In addition to the terms discussed above, most agreements will also contain a fair quantity of legal 'boiler plate' that deals with issues such as:

- disclaimers covering the usefulness and fitness for purpose of background IP, materials and so on,
- force majeure stating that partners will not be responsible for failures to deliver its obligations if the reasons are beyond their reasonable control,
- non-assignment, in other words membership of the consortium cannot be transferred or re-assigned to a third party,
- continuing obligations that survive the duration of the project such as confidentiality and IP licenses,
- the governing law of the contract
- no partnership, in other words the project is not a joint venture or partnership as described by company law, but rather a collaboration, and
- dispute resolution measures that control how disagreements between the partners are dealt with.

2.3 Working with lawyers

As will have become obvious, professional legal assistance will be required to agree and draft the consortium agreement. It is perhaps not best practice however to leave the negotiations entirely in the hands of each partner's legal representation. The results are likely to be expensive and not quite what the partners envisaged as, quite rightly, each partner's lawyer will concentrate on the rights of their client rather than thinking of the greater good of the project as a whole.

It is therefore recommended that the partners agree the basic terms and heads of agreement first, concentrating on the key aspects of intellectual property ownership and use, publications and project management. The lawyers can then draft up the agreement around these principles and add in the necessary 'boiler plate' to protect their clients.

As I warned you, not the most exciting of topics, but hopefully you will appreciate its importance and the need to consider the terms early on in the development of your plans. Well done, I think you have earned another biscuit!

2.4 So in conclusion

1. All collaborative projects should be covered by a consortium agreement that defines:
- the scope and definition of the project
- the responsibilities and liabilities
- how the project will be managed
- the role and structure of the Steering Committee
- changes to the consortium
- termination of the project
- financial management
- intellectual property rights
- publications
- confidentiality
- legal boiler plate

2. Professional legal assistance should be sought when preparing and writing collaboration agreements.

3 Getting off to a good start

Managing a collaborative project differs considerably from normal project management for two reasons. Firstly the objectives of the project are multiple, in other words not all the participants share the same motivations for being involved in the project. Secondly, not all the participants work for you. In a traditional project, even if you use sub-contractors, the project manager normally has some line management control over all the resources involved in the project. In a collaborative project they do not.

We are going to look in detail at two key areas that need to be addressed before the project starts. The first is the development of a business case which clearly explains the commercial reasons for carrying out the project. This could be used to win the backing of each partner's senior management or form the basis of your grant applications.

Secondly we are going to look at project planning within a collaborative context. Whilst it is not the intention of this book to cover the topic of project management, I will share with you a highly effective project planning method that you can use to build a project plan that the whole consortium understands, is bought into and is committed to achieving.

3.1 The business case

The purpose of the business case is to clearly define the justification for carrying out the project. The justification is based on the commercial or other opportunities it will create, balanced against the costs and risks involved in carrying it out. The business case is therefore used to say why the expected time and effort will be worth while. The project steering committee can use the business case to regularly monitor the ongoing viability of the project. The business case document therefore, like so many others in project management, becomes a living document that is referred to and revised continually throughout the duration of the project.

The first thing to define in your project's business case is the set of objectives that you and your partners want to achieve. As already mentioned, the objectives of collaborative projects are often multiple. There will be a principal objective such as to build a prototype that proves that a certain technology can provide benefits to customers, or to solve a technical barrier that is preventing the exploitation of a new technique. However there will be other individual objectives such as to provide an opportunity for a small business to work with a larger player in the market, to improve the efficiency of your production or marketing process, to see the results of your research work being applied and so on. Ideally everybody in the consortium should understand everyone else's objectives up front so that the project plan can be built to satisfy everyone.

There is another reason why it good to air everyone's objectives up front. In some cases, individual objectives may be incompatible with each other. This is best discovered at the very beginning rather than later on when real money has been committed. Not all incompatibilities are show stoppers so the earlier conflicts are recognised, the easier they are to deal with and accommodate. An obvious and common example is the desire of academic partners to publish the results of their work and the desire of a commercial partner to retain intellectual property and trade secrets. By airing the objectives at the very start, the terms of the consortium agreement can be drafted to accommodate both wishes and the project's intellectual property protection policy can be designed to protect new innovations before publications are released.

> **Stakeholders' workshop**
>
> Whilst it is possible to collect individual partner's objectives by email or by the project manager talking to everyone individually, it is a good idea to bring all the project team together to talk about their objectives. Getting all the stakeholders together allows discussions to take place straight away to overcome any conflicts.
>
> You should try to bring senior management and marketing executives along to this meeting so that they understand what you are trying to achieve and so they can agree on any compromises without you having to continually refer back to the office. This also allows the technologists to understand the commercial reasons for carrying out the project and for the business people to understand what might be technically possible as a result of the project.
>
> It may not be worthwhile holding a meeting solely to talk about objectives, but it should certainly be an agenda item at an early scoping or planning meeting.

Qualitative and quantitative objectives

The objectives that you and your partners have for carrying out a collaborative project will fall into two categories. Those that have no direct or tangible financial benefit are qualitative objectives. Those that are measurable in terms of a financial return are quantitative.

Qualitative objectives might include:
- the opportunity to work with a larger company in the market,
- the opportunity to disseminate research results to a wider audience,
- improving the public image of the organisation by supporting a worthy project,

- investigating a new potential market space or supply chain,
- reducing the environmental impact of a product or process,
- carrying out or supporting basic research, or
- developing skills or providing new career development opportunities for staff.

Quantitative objectives might include:
- developing a new product or service,
- improving the efficiency of a production or service delivery process,
- assessing the potential of a new technology, or
- developing new licence opportunities.

When developing the business case for a project, it is necessary to understand the worth of the objectives, so that they can be balanced against the costs and risks of carrying out the project. The qualitative objectives, having no direct financial return, are very subjective and each organisation will need to find their own way of justifying the spending involved in achieving them.

In justifying the quantitative objectives however, we can be a little more systematic.

Analysing the benefits

For each of the project's quantitative objectives, a business analysis needs to be carried out. You need to know who the customers are, what problems you are going to solve for them and what scale of return you can expect on your investment. This analysis will be repeated and gone into in more depth as the project proceeds and as you develop your exploitation strategies and plans. For now though, you need enough of an idea to give everyone the confidence to proceed.

At this stage we are interested in the widest possible set of opportunities that the project can open up. This means that we need to analyse all the possible customer groups for each of the project's commercial partners.

Define the customers

The first step is to create a list of the potential customer groups that could benefit from the outcomes of the project. It doesn't matter if none of the partners can access a particular group, because this might lead to new licence opportunities. A little research will be required for each set of customers you define. You need to know how large a group they are at both the national and global level and if possible what they spend on products or services of the type you intend to develop.

This list of potential customers could be used to form a set of use cases. We will look in more detail at use cases in chapter 8.

Customers are often referred to as end-users in R&D project circles and up to now I have referred them this way. I do find however, that it is better to think of them from the outset as customers as it helps to focus your mind on the fact that they have a choice whether to use the results of your project or not. The more you think of them as customers the more you accept that you have to sell to them and satisfy them in order to realise a financial return from your project.

Analysing the problems to be solved

At first glance, this heading might make you think about the technical issues that need to be overcome if you are to successfully meet the objectives of the project. This is not however the case. For now we need to analyse the problems faced by your customers, which the project outcomes will overcome for them.

All commercial products and services rely on customers handing over their hard earned cash to suppliers in return for them solving a problem for them. This simple rule applies to every quantitative objective that your project is likely to have. To achieve the best results, you need to understand how big a problem you are addressing and therefore how much the customer is going to be willing to pay you to overcome it.

Imagine a project that could improve the efficiency of a production process to the extent that raw material wastage is reduced by 50%. The problem being addressed is that the current process results in material waste. However to understand how big a problem we are addressing we need to know what proportion of material use is currently wasted and how much this costs. If the current system results in 20% waste and we are going to reduce this to 10%, this will reduce material costs by 8%. However if current wastage is just 2% and we reduce it to 1%, this will only reduce material costs by 0.8%. The scale of the problem is quite different. The value of the material wasted is also a factor. If the annual material costs are $1million, the 8% saving equates to $80,000 per year, but only $8,000 per year for the 0.8% saving. If however the materials costs are $10million per year, the 0.8% saving now provides an $80,000 return. If the project costs $10,000 to run, it is still very worth while carrying it out.

For each of your potential customer groups, define what problem you are going to able to solve for them and how great a problem it is to them. In many cases you will be able to use empathy or your team's knowledge of the customer base, but much better results can be achieved through talking to them face to face.

Talking to customers

Talking to the project's potential customers can have many rewards. Firstly it is an excellent way of understanding the problem you want to solve and their demand for it. It can also be an excellent opportunity for your sales and marketing teams to get involved with the project and bring their thoughts to the table. It shows your customers that you care about them and are advancing your products and technology in exciting areas.

Customers that show a keen interest in the project and its results should be kept up to date with the progress of the project, either informally or through the creation of an advisory or user group.

It is not uncommon for customers to want to join a good project that addresses their needs. This is an excellent result if you can achieve it. End user partners often provide valuable resources such as real data, materials or testing opportunities. They also provide a ready market and testimonials.

The market opportunities

Now that you have a list of potential customer groups, you need to find out how many customers there are in each. This is simple market research and in many cases the commercial partners will have this information to hand already. At this stage we are not trying to estimate sales or market penetration, we are simply trying to measure the opportunities that the project could open up. For example if one of the customer groups is UK general medical practitioners, you need to know that there are 32,000 of them, arranged into 152 purchasing groups.

This analysis is very important in shaping the direction of the project towards the best opportunities. We need to know the scale of each opportunity in order to aim the project towards the ones with the best potential return.

The dynamics of each potential market also need investigating. Try to establish how the market has changed over the last five years and how experts, such as market analysts, predict it is going to behave in the future. If the market is forecast to decline, there may be little point in you carrying out a project that tries to exploit it.

You also need to consider how accessible each of the opportunities is for the partnership. Opportunities in existing markets will be much more easily addressed than ones in new and unfamiliar areas.

Finally, rank the opportunities by their attractiveness based on their size, level of customer demand, likely future growth and accessibility. The final list should contain at least one target customer group for each commercial partner within the project.

The project deliverables

Now that we know a little more about the intended target customers, it is possible to define the ideal outcomes and deliverables that the project should achieve in order to best satisfy them.

By coming at the project plan from this direction, you define what you want from the project from the perspective of the target customers. This can often lead to a project that is quite different from your first thoughts, but one that will be better received and potentially much more worthwhile.

Potential returns

We now need to put a value on the benefits of the project so that it can be compared to the costs and risks involved in carrying it out. We know the potential market sizes; we now need the price, costs and the penetration rate.

There is no rocket science involved here, we simply need to guess what the customer would be willing to pay and then guess what it will cost to deliver. The difference is gross profit per sale. Get the best advice you can for these estimates, ideally from the sales and marketing teams as well as the technologists from the commercial partners. If you have an advisory group or 'tame' customers ask them what they would expect to pay.

Now we have the market size and the unit profit, but simply multiplying them together would yield an unrealistic total. A sensible penetration rate is required. The penetration rate is the proportion of the total available market that could reasonably be expected to buy from you. Again this is a guess, but rather than just say 10%, it is better to use a forecasting methodology such as the AIDA model illustrated in figure 3.1.

AIDA is an abbreviation of:
- Awareness – what proportion of the market can you make aware of your product.
- Interest – of these, how many will be interested in your product.
- Decision – of these, how many will decide to buy it.
- Achievement – of these, how many will you actually achieve a sale from.

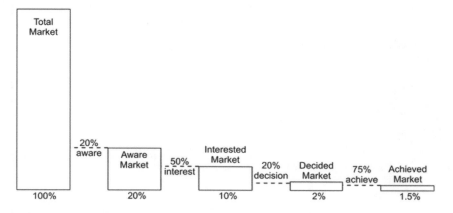

Fig. 3.1. The AIDA model of customer penetration

Taking the UK GP market as an example, imagine a new diagnostic tool. The total market potential is 32,000, if we say we can make

20% aware of it, that's 6,400. Of these, 50% will be interested, that's 3,200. Of these, 20% will decide to buy it, that's 640. Of these, we will achieve 75% satisfactory completion, that's 480. The penetration is therefore 480/32,000 or 1.5% as illustrated in Figure 3.1 above.

If we estimate that we could sell the product for $2,000 and it will cost us $1,000 to make, then the potential return is $480,000 for this market place.

Each partner will need to do their own calculations to estimate what returns could be possible for them. Each partner then needs to compare the potential return with the costs of being involved in the project.

Potential return pitfalls

There are a number of common mistakes often made by grant applicants in estimating potential returns. Try to avoid the following top 3 mistakes:

1. The returns are based on market sizes measured financially rather than in potential sales or numbers of customers. 'This is a $10billion market, if we only get 1% we will be rich'. You should estimate market penetration through numbers of customers rather than these broad numbers that often include vast areas that your product or service will not compete in.

2. Not knowing the market numbers at all. This could be a fatal problem for your project. If your consortium does not know or understand the market they are trying to exploit, you do not have the right consortium.

3. The project will speed up a process, such as testing, thus saving a large amount of money for the customer. Firstly the potential saving does not equate to the price someone will pay to make the saving. Secondly make sure your customer

> wants to save money, they may prefer to carry out more tests from the same budget. You still have a competitive advantage, but the selling point is test speed rather than cost reduction. This will influence the approach your project takes.

If it is likely to cost more to run the project than the return that can be generated, then the project is not worth doing for the quantitative objectives alone.

The project risk also needs to be taken into account. If the project is of high risk, a simple return greater than the costs is not sufficient to justify the project. A risk premium needs to be applied to the project costs to produce a higher justification threshold. For example, if the project costs $100,000 but is of low risk, you could apply a 50% risk premium so that the return would need to be greater than 1.5 times the project costs, or $150,000, before it could be justified. High risk projects might have risk premiums of 200 or even 500%. You will need to ask senior management what kind of return they will accept to justify your project.

You may also want to take into account that the return on investment is based on future sales. Accounting techniques such as Net Present Value (NPV) calculations may be used to work out the current value of future sales so that this value is used in the justification.

Do you have a project?

At this point it is worth taking stock and asking yourself if you still have a project. In order to proceed you and your partners need to be convinced that there is a market demand for the project outcomes, there is a significant and achievable financial opportunity and that all your other objectives can be met.

At this point however, you don't really know how much the project will cost to carry out, so the next step is to create a project plan and find out.

3.2 Collaborative project planning

Within any project, getting the commitment and enthusiasm of all the participants is essential. When people are highly motivated, they work hard to deliver on time, on budget and to scope. Within collaborative projects it is even more important because the work is normally distributed between the partners, who can frequently be hundreds or even thousands of miles apart. The project manager is not able to monitor the project quite as closely as if they were all working on the same site. Holding weekly face to face meetings is seldom possible, so the project manager must therefore largely trust each partner to be following the plan.

It is therefore essential that the plan has the commitment and understanding of every member of the consortium. This cannot be achieved by having a project manager devise a plan in isolation and then to impose it on everyone else. The process of creating the project plan has to involve the whole team so that it is understood by everyone and most importantly 'bought in to' by the whole consortium.

The following methodology can be used to achieve this. The process can easily be achieved in half a day for most projects, or a whole day for large or complex projects.

First, arrange a date on which the whole consortium can attend a planning workshop. Encourage each partner to bring the key technical people that will be working on the project, but try to keep the total number of delegates to a maximum of 15 or so. If the project is grant funded and a monitoring officer has been appointed, you might consider inviting them along to help, provide advice and get to know the team. To run the workshop, you will need plenty of space, at least one large white board, lots of stickies (such as Post-it® pads), pens, a digital camera and of course a good supply of coffee and chocolate biscuits.

The objectives and business case should be well understood by everyone at this stage so the task for the day is mainly to develop a project plan that will satisfy those objectives. It might however be useful to set the scene by reminding everyone of the project and partners' objectives.

You will need to appoint a facilitator for the workshop. Ideally this should be the project manager who will be responsible for the day to day coordination of the project and reporting to the steering committee.

Step 1

The first step is to generate a list of all the tasks that need to be carried out during the project. This is achieved by the whole group brainstorming and calling out tasks as they occur to them. The facilitator writes the task on a sticky and slaps it onto the white board. The position is not important at this stage, just that it is recorded.

As the facilitator, lead the generation of tasks through each area of the project, pose questions such as 'do we need to...', 'how will we...' and so on. Don't forget areas like market research and project management.

When the board is getting full and the flow of suggestions is starting to dry up, stop. Don't worry if there are gaps, these will get filled in very quickly in the following few steps.

Your white board should resemble figure 3.2, a mass of unstructured tasks that all need to take place during the project.

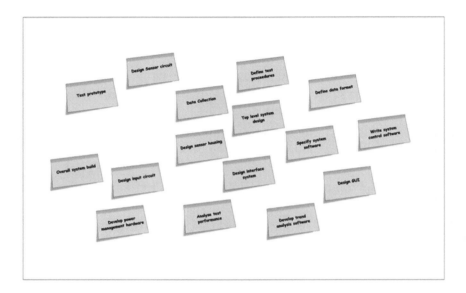

Fig. 3.2. The board after step 1

Step 2

Next, arrange the stickies into groups that describe the main areas of work within the project. The objective is to try and arrange them into a small number of groups, between 4 and 8 is ideal, although some may have sub-groupings. Again the activity is led by the facilitator. You can write the area of work headings down the left hand side of the white board and move the stickies into the correct row.

Involve the whole team in the definition of each area of work and what tasks go where on the board. If new tasks are thought of, write them on a sticky and add them to the appropriate row.

This is quite a rapid step as the separation of tasks into these areas of work normally becomes very obvious. When the step is complete, your white board should resemble figure 3.3.

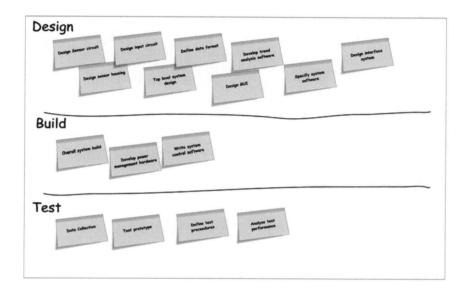

Fig. 3.3. The board after step 2

Step 3

Within each area of work, arrange the task stickies into the correct logical chronological order. Here again, new tasks are likely to be generated as the gaps between the existing tasks are recognised.

Work through each area of work, one at a time, still involving the whole group, even if some of them are not likely to be involved in all the areas of work. It is important that they understand what else will be going on within the project.

Again, this step is fairly rapid. Figure 3.4 shows the state your board should be in by now.

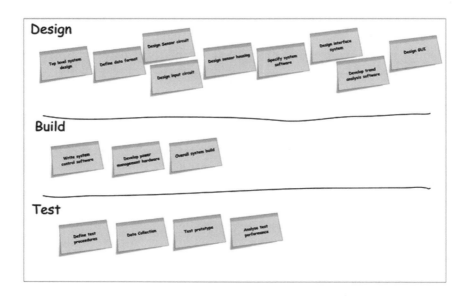

Fig. 3.4. The board after step 3

Step 4

The next step is to define the dependencies between each of the tasks. These will exist both within and across each of the areas of work. Dependencies between tasks exist when, for example a task cannot be started until a preceding task is completed.

There are four basic types of dependency:
- Finish to Start – this is the most common dependency and means that a preceding task has to be finished before the next task can start. For example the painting of components must have been finished before assembly can start.
- Start to Start – this dependency states that two or more tasks must start together, this is necessary when, for example two testing tasks use a common testing environment.
- Finish to Finish – this applies to two or more tasks that must finish together. For example the preparation of perishable specimens that are to be analysed together. The analysis cannot start until they are both ready but one cannot be

allowed to deteriorate waiting for the other to be prepared. The two preparation tasks must finish together.

- Start to Finish – this dependency is employed when a task cannot complete until another has started, akin to passing a baton in a race, you can't stop running until the next team member has taken the baton from you.

For each task, work through what, if any, dependencies exist with other tasks, both in the same area of work and in the others. These can be drawn onto the board between the stickies. If there is space, move the stickies from left to right so that they start to line up with each other according to their dependencies.

Again, more tasks may be generated at this stage and these should be added as appropriate. Now is also a time to rationalise some of the tasks, to combine very close tasks into one and so on. When this happens, replace the two old stickies with a new one.

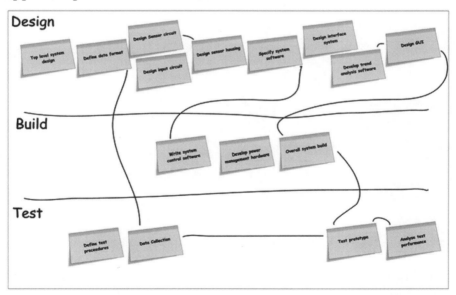

Fig. 3.5. The board after step 4

The result is starting to look like a project plan. It is not a full Gantt chart just yet though because there is no duration information known about any of the tasks. What you have created though is the project's structural chart. It should resemble figure 3.5.

Step 5

Now that we have a fairly stable set of tasks, it's time to number them and take some notes. Firstly, allocate letters to each of the areas of work. You could simply use A,B,C and so on, but it might make more sense to use the task initial, for example if the areas of work were Design, Build and Test, you could use D,B and T.

Next allocate a number to each task in their logical order, starting with 1 for each area of work, so that tasks are numbered D1, D2, D3, B1, B2 and so on. Write the relevant task number in the top left hand corner of each of the stickies.

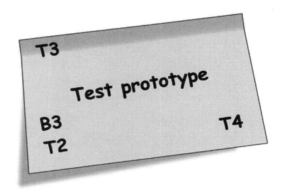

Fig. 3.6. A typical stickie showing its task number and dependencies

Finally in the bottom left hand corner of each sticky, write the task numbers of any tasks that this one is dependent upon, and in the bottom right corner, write the task numbers of any tasks that are dependent upon this one. Each sticky should look like figure 3.6.

Step 5.5

By this time everybody will need a break. Pause for a while and allow everyone to take a comfort break, refill their mugs, check their messages and so on. Before the next step, take a few photographs of the board.

Step 6

The next step is to work in detail on each of the areas of work. It may be possible to split the group up to work on each area in parallel, but only if sufficient representation is available from each partner to cover each of the areas they are going to be involved in.

Each task sticky now needs to be further defined into a work package description. To do this additional information is required about it such as:
- What the work package objectives are,
- What the deliverables or outcomes of the work package will be,
- What each partner's contribution to the work package will be,
- How much work is required and what resources will be made available,
- How long the work package will take to complete,
- What equipment, facilities or materials will be required,
- Who will take responsibility for the work package,
- What the main technical risks are.

Working through the work packages one by one, define each one using a form like the one shown in figure 3.7 to capture the information. You may find that tasks merge at this stage to become larger work packages; this is fine as long as the work packages do not become too big. As a guide, if their resulting duration becomes longer than 6 months or their total effort becomes more than one person-year, they should not be merged. When tasks are merged,

Work Package Title:		Number:	
Work Package Leader:			
Contributing Partners:			
Work Package Objectives: ◊ ◊ ◊			
Description of Work (who will do what):			
Deliverables and Outcomes: ◊ ◊ ◊			
Equipment and Facilities Required: ◊ ◊ ◊			
Effort Required by: Partner Name	Effort in Man Days per Role Required		
Work Package Duration:		Total Effort:	
Technical Risks: ◊ ◊ ◊			

Fig. 3.7. Work package description template

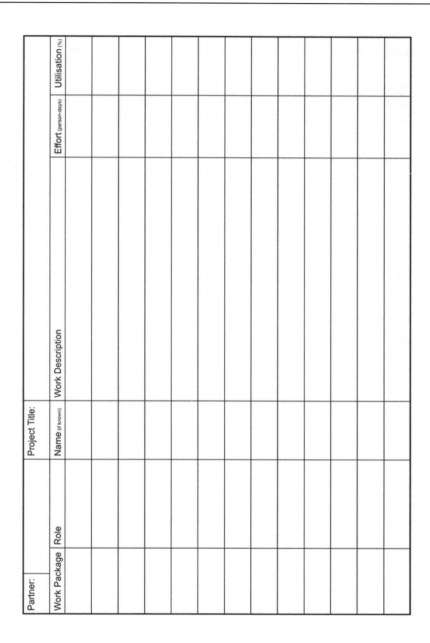

Fig. 3.8. Partner commitment form template

replace the original stickies on the board with the new combined one, giving it the number of the first removed task.

In addition each partner should keep a tally of the resources they commit to each work package using a commitment form like the one in figure 3.8. If the resource is labour, note the effort committed in person days, if the role is not full time, add a utilisation factor to help calculate the actual costs.

Step 6.5

Time for another break, you can reassure everyone that they are nearly there now!

Step 7

Now that the duration and resource requirements are known for each work package, you can convert the structural chart into the final Gantt chart.

Depending on the facilitator's skills with project management software, this next step could be done using a computer and LCD projector or you could carry on using the white board and do the IT bit later on.

Start with a fresh white board. Remove all the stickies carefully, you have not finished with them yet, and clean the dependence lines. Mark new time lines as vertical lines down the whole length of the board, use quarter years and add as many as you think you will need.

Next bring the whole group together and work through each area of work, repositioning each sticky on the time line in the position that the work package should start. Indicate the duration of the task as a horizontal line of the appropriate length immediately below the sticky. Draw in dependencies between the tasks as they occur. The finished plan will look something like figure 3.9.

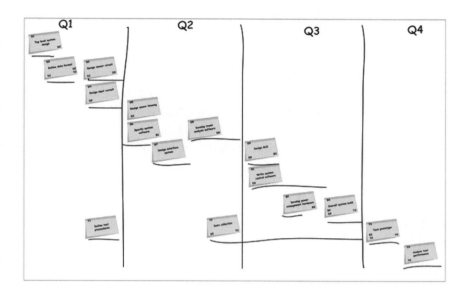

Fig. 3.9. The board after step 7

When you have finished, the resulting chart is a potential Gantt chart for your project. However there is still one more step before you have completely finished.

Step 8

The resulting plan from the previous step will only work if there are sufficient resources and people available to carry out all the tasks at the times they are shown. The final step is to examine what resources are required for each task and adjust their timing to smooth out the peeks and troughs in overall effort. In project management circles this is known as resource levelling and seeks to ensure that, for example you are not committed 200% for this week, but idle the following week.

Work across the plan from the start. For each month or quarter depending on the resolution of your plan, let each partner add up their commitment to the active work packages using their commitment forms. Where resources are over committed, working as a group, move the work packages back and forth, within their dependence constraints, to smooth out the issues.

Continue until everyone is happy with the plan. This may not be easy and in some cases you will have to extend the duration of the whole project to achieve a plan that is workable.

Everyone can now relax a bit, except for the facilitator who must now collect all the work package definitions and prepare to convert them and the contents of the board into a formal project plan document. Take plenty of photographs of the board to make sure you don't lose any of the vital links and positions. Check that they can be read clearly before wiping the board though!

Step 9

I know I said step 8 would be the last one, but actually the last step is to congratulate everyone for bringing together a plan that is thoroughly thought out, that everyone in the team understands and most importantly, a plan that everybody in the team believes in and is committed to.

Take a team photo too.

Follow-up work

It is important to get all the information into a formal plan their approval. However, there may be additional information required from each partner to complete this work.

A spend profile is very often required by funding bodies to visualise the budget of the project and the commitment being put into each work package. Even if a funding body is not involved, a project budget shows everyone who is spending what within the project.

Creating the spend profile is relatively straight forward. Each partner has their commitment form which shows what resources they expect to commit on each of the work packages they are involved in. They also have a copy of the project Gantt chart which shows when each work package starts and how long it will last. The spend profile simply adds all the active work package spends in each month or quarter to create a total commitment for each period.

Once the pure work package commitments are done, add in additional expenses such as attendance at project meetings, reporting, equipment and material purchases and so on, to create the finished spend profile. Use a spreadsheet to enter all the data. The project manager should create a template for each partner to complete so that all the data will be compatible when it is collated.

The project manager should collect the spend profiles from each of the partners to create a master spend profile for the entire project. This can then become the project budget and be used to monitor project spend as the work progresses.

Do you still have a project?

When you completed the business case, you knew what the potential return on the project could be. You now know what the project will cost in terms of time and resources. You can now compare the numbers, apply a sensible risk premium and see if you can justify the project.

If the project is justifiable on the quantitative objectives alone then this is great news. If not, each partner will need to consider the worth of their qualitative objectives in order to fully justify their involvement in the project.

If applicable, grant contributions should be included in the justification calculations. If you are expecting a grant contribution of 50%, reduce your project costs by 50% before applying the risk premium and comparing it to the justification threshold. These calculations could form part of the additionality case for your grant application. For example, 'to justify a project with this level of risk, we would need to see a return of 300% on our investment, this is only possible with grant support equal to 35% of project costs.'

There you have it, a sound business case and a complete project plan, now all you have to do is make sure everyone follows it!

3.3 So in conclusion

1. Two key pieces of work are required before the project can start. These are the creation of:
- A business case
- A project plan

2. The business case should include an analysis of:
- The objectives of the partners
- The benefits expected from the project
- The customers or end-users of the project outcomes
- The problems solved by the project outcomes
- The market opportunities
- The desired project deliverables
- The potential returns on investment

3. Within collaborative projects, the project plan must be:
- Understood by everyone in the consortium
- Bought in to by each partner
- Committed to by each partner

4. The project plan can be created in a workshop using the following steps:
- Generate a list of tasks
- Define the principal areas of work

- Ordering the tasks within each area
- Defining the interdependencies between the tasks
- Numbering the tasks
- Carrying out detail planning of each task
- Defining the duration and effort required for each task
- Creating and smoothing a Gantt chart
- Creating a spend profile

4 Maintaining momentum

This chapter is intended to help you run your project. I have already said that I don't intend this book to cover the basics of project management; there are any numbers of excellent resources to help you with that. There are however, some aspects of collaborative projects that need special attention.

Firstly, your project collaborators are coming together as a coalition of equals. This means that you cannot boss them about in the way that you might with sub-contractors. You will need to establish a democracy where each company's voice is heard and you will need to meet regularly to monitor and control the progress.

Detailed planning is also very important in collaborative projects where the work is distributed and keeping to time and budget are essential.

Risk management is a key part of managing any project, but this is one area that is often neglected in collaborative R&D projects.

Leading a collaborative project is a non-trivial exercise. Whoever takes overall responsibility for the management of the project will require a special skill set and personality.

Finally, if your project is supported by outside funding such as a grant, the chances are you will have a monitoring officer put into place to monitor progress on behalf of the funding body. The project manager will need to work closely with the monitoring officer. If this relationship is well managed the monitor can often add good value to the project.

4.1 Project control

Figure 4.1 shows a typical management structure for a collaborative project. The steering committee takes overall responsibility for the direction of the project, but the day to day management is delegated to a project manager. The individual work package leaders all report to the project manager. In addition various sub-committees can be established to look at specific issues such as exploitation, publication approval and so on.

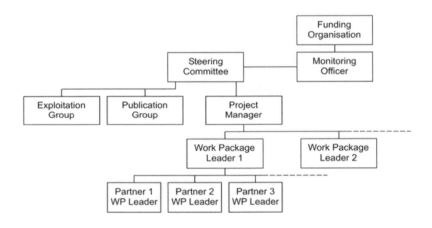

Fig. 4.1. Collaborative project management structure

The project manager will need to impose a reporting regime on each of the work packages. In the case of many grant funded projects, reporting templates are imposed on the project by the funding body. If not, the project manager will need to devise a reporting system which includes the following basic elements:

- A short statement about the progress and achievements made,
- A statement about the progress towards the relevant deliverables and milestones,
- A review and update of the risks involved in the work package,

- A description of any issues that have arisen,
- A statement of the resources and costs expended to this point within the work package and
- An update of the estimated remaining time and resources required to complete the work package.

It is important to keep the length of these reports as short as possible, ideally to a single page per work package. You need the work package leaders to be working on the project rather than spending all their time writing reports. However good communication is a vital part of project management and is essential in collaborative projects where other means of communication between the different partners may not exist.

The project manager should request reports from each work package at regular intervals and prior to each steering committee meeting. Depending on the scale of the project, you might expect there to be a series of work package meetings to agree the contents of the reports before they are submitted to the project manager.

Steering committees

Like any democratic society, your collaborative project will need to be run through a series of committee meetings. At the top of the system there should be a Steering Committee that takes overall responsibility for the project.

The steering committee's role is similar to that of a board of directors in a company. The committee appoints the project manager and other executive roles within the project, approves the project plans, monitors progress against them and allocates the resources required to achieve the project objectives. In addition the steering committee has the right to remove partners from the consortium and invite new partners to join.

The Consortium Agreement document should define the structure and voting rights of the committee. Typically this will include the right for each consortium member to have one seat on the committee, with one vote per seat.

Putting it to the vote

Although the steering committee has voting rights, being forced to put issues to the vote is a sign that all is not well within your project. It is far better for all the parties to reach agreement on all the issues through discussion and negotiation. This will lead to a much more supportive and collaborative environment.

The only time when you might want to formally take a vote is when dealing with the removal or addition of partners within the consortium, the vote being recorded in the minutes.

The steering committee should meet regularly throughout the project. Depending on the scale and duration of the project, this might mean monthly or quarterly. You can of course increase the frequency during times of increased activity or importance and any member should have the right to call a steering committee meeting if they have a burning issue that they consider needs urgent attention.

Steering committee meetings should not be rushed. In many cases, representatives have to travel to reach the meeting and will want to spend time understanding how the project is progressing, especially in the areas in which they are not directly involved. A typical meeting might start at 10am to allow people to arrive in time and conclude by mid afternoon. Allow plenty of time for lunch as this is a great opportunity for people to chat informally about the project and other opportunities that might be relevant.

It is also a good idea to rotate the venue of the steering committee meetings around the project partners. This allows each partner to act as host and for everyone to get a chance to look around each others facilities. It is also a good idea to agree the dates and venue a long time in advance as getting everybody's diaries together with a week to go is unlikely to work. Planning the dates of meetings a year ahead can save you a lot of trouble.

The agenda for the steering meeting is of course up to you. You will probably want to include the usual items to review previous minutes, discuss matters arising and receive updates from each of the work package leaders. However you might also like to consider the following as standard agenda items:

- A review of the status of milestones and deliverables. If you present these as a complete list and colour code them so that, for example, late items are shown red, completed items as green and imminent items as blue, it is easy for everyone to quickly appreciate how the project is progressing.
- A review of the risk register,
- A review of project spending and resources,
- A review of the use and generation of intellectual property. This could include updates on the use of background IP as well as the notification of new IP and its ownership,
- A review of project publications and dissemination,
- An update of the project business case and the development of exploitation plans,
- If the project has a monitoring officer, ask them for feedback on how they see the progress of the project.

The use of logs

Logs are an excellent way of keeping track of important information that will save a lot of time and effort at the end of the project. An IP log for example that records the use of background IP and the creation of new foreground IP will greatly aid the licensing negotiations when the comm- ercialisation of the results is planned.

> The log document can be a simple table but all partners
> should have read access to it. The project manager should be
> responsible for adding new entries on notification by the
> partners.

4.2 Detailed planning

The project plan created in the last chapter is a top level view of the
project. It tells the steering committee what is happening and when
within the project. Unless the work packages are very simple, there
may be a need to carry out more detailed planning within each
individual work package.

Fortunately exactly the same process can be used to create work
package plans. Just before the work package is due to start, the
team assembles and works though all the individual tasks that are
required to deliver the work package. The same templates can be
used as task descriptions as were previously used for work package
descriptions.

In order to maximize the benefit of this detailed planning process, it
should be carried out as each new work package starts. This way
the work package can be planned using the very latest information
from the preceding work and the most accurate estimates of
realistic work rates. This is known as 'rolling wave' planning as the
plans develop as the project rolls forward.

Some people complain that they don't have time to carry out all this
planning and claim it is unnecessary, however they often seem to be
able to find the time to do the work twice when it goes wrong!

4.3 Managing risk

The management of risk is an important part of any project manager's role but in collaborative R&D projects, the assessment, avoidance and mitigation of risks are even more important.

Two aspects of collaborative R&D projects add significantly to their risk. Firstly they are collaborative and secondly they involve R&D. This may seem a bit flippant and obvious but traditionally the management of risk is an area that is often at best neglected, and at worst ignored all together in these types of project.

Types of risk

When we think of risks, we tend to think about things going wrong, however within projects, risks can be divided into operational risks that have an up as well as a down side, and insurable risks that only have a downside.

Operational risks

Most risks within a project will be operational risks. These include:
- the performance of technology
- the availability of resources
- response of the market to the new idea
- and sometimes the environment or weather

The expectation is that the affects of the risk will more often be negative rather than positive. For example the impact of the lowest possible response of the market to the product will be greater then the most positive. Weather can act in the same way. Weather could stop field work or even destroy previously installed equipment but is unlikely to be good enough to allow you to proceed at twice the expected rate.

The project plan should reflect the best guess of how operational risks are most likely to play out but include contingency plans for deviations from this position.

Insurable risks

Insurable risks lead to downside only. They are called insurable but they are not always covered by the partners' insurance policies. Insurable risks include:

- property or facility damage
- consequential losses
- legal liability issues
- personal loss or injury

Property damage may be caused by fire, flood or damage caused to equipment during transportation or delivery. Consequential losses include lost time due to facilities being unavailable or through staff illness. Legal liabilities may arise from loss or damage of property, negligence on the part of suppliers, or injury to a third party. Liabilities may also arise from contractual problems between the partners. Finally there are the risks associated with members of the team or general public being injured or having property damaged through the running of experiments or trials.

Risk management

Risk management is concerned with the identification of risks and minimizing their impact on the project. There are four basic steps involved:

1. identification of the sources of risk within the project
2. determination of the impact and likelihood of the risks occurring
3. developing strategies for forecasting risk likelihood
4. developing strategies for mitigating the affects of the risks should they occur

Identification of risks

Unfortunately this is not a time to be optimistic. You need to identify all the things that can go wrong with the project. Hopefully the work package leaders will have made a good start and will have identified the main technical risks that are associated with each work package. As a group you will also need to look at the technical risks associated with the transfer of work from work-package to work package.

The technical risks are perhaps the easiest to consider, especially when the project is technical in nature. However in addition, there are also commercial risks, managerial risks and environmental risks.

Commercial risks may include shifting market dynamics or the response of the customers to your ideas. For example what would your strategy be if your user group's response to the ideas was poor or even critical? How would you manage the situation?

Managerial risks might include staff or resource availability issues, communication issues between the partners or issues involved in making grant claims.

Environmental risks might include impacts caused by climate or pollution which affects your work or it may involve your project creating environmental or social problems for others. How would you avoid or deal with complaints about your field trials that involved erecting semi-permanent structures or created a disturbance or nuisance for someone else?

When it comes to identifying the list of risks, a number of techniques are available. The simplest form of risk identification is simply to use your expert judgement, your intuition and experience. This may be sufficient for simple projects but for more complex situations a more structured approach will be required.

Plan decomposition is a technique that involves the examination of all the deliverables, work package starts and ends and the milestones that indicate the bringing together of a number of project strands. The risks associated with successfully achieving each point are then considered.

Both these techniques are greatly enhanced through brainstorming with the whole team.

Impact analysis

The purpose of impact analysis is to calculate a score for each of your identified risks so that they can be rank ordered and addressed in priority order.

The impact of a risk is often calculated by multiplying together a score for the likelihood of the risk occurring and the consequence of the risk to the project. This has the benefit that risks that are highly unlikely or of no real consequence are demoted where as risks that are both likely and damaging can be given priority. For example the risk of a meteor striking your laboratory would score highly for impact (sorry!) but very lowly for likelihood and would therefore not require further thought. Delivery delays for specialist equipment purchases might score highly on both counts and would therefore be more worth worrying about.

> **Scoring risk**
>
> The likelihood and impact should both be scored out of 5 resulting in risk scores between 1 and 25. This should be sufficient to provide a sensible rank order of risk. A range of 5 is also easy to discuss in terms of very low to very high (VL/L/M/H/VH).
>
> The scores for each risk should be reassessed at every reporting period and the top 10 discussed at the steering committee meeting.

Forecasting risk

It is useful to extend the risk likelihood analysis to include a forecasting strategy. Forecasting implies that you have some method of predicting whether a risk is more or less likely in order to provide a warning and an opportunity to instigate your mitigation strategy.

Imagine you need to erect an instrument tower that requires planning permission. There is a risk that planning permission may be refused. The impact on the project would be high, let's say 4. The likelihood is medium so 3. This gives a score of 12 which is worthy of consideration. A simple forecasting strategy would involve discussing the case with the authorities and keeping in contact with them throughout the application process. That way you can gauge your chances of success as the application proceeds.

Mitigating risk

There are three types of risk mitigation strategy:
- Avoidance
- Deflection
- Contingency

By way of example, imagine a sensitive field instrument that doesn't like exposure to the rain. If it is in place and bad weather is forecast you have three choices. You can bring it indoors (avoidance), cover it up (deflection) or leave it where it is and fix it if it does get damaged (contingency).

Avoidance strategies are those that involve you re-planning aspects of the project so that the risks are never realized. Such strategies are required where the impact is very high or even fatal to the project. You may be able to develop avoidance strategies that although do not reduce the impact, do reduce the likelihood thus lowering the overall risk score.

Deflection is a common technique in traditional project management and involves deflecting the risk outside of the project either to an insurance provider or some other third party, typically the client. Deflection strategies are therefore less common in collaborative R&D projects where there are limited opportunities for 'passing the buck'.

Contingency planning is perhaps more appropriate to R&D projects than traditional projects. Traditional projects prefer to avoid risks by using well understood technologies and processes. R&D projects by their very nature involve use of unknown technologies and processes and treat the resultant risks are wholly acceptable.

The approach simply involves creating alternative, or contingency, plans that are actioned should a risk occur. This can either be carried out at the work package level or at the project level depending on the risk.

4.4 Project leadership

The task of leading a collaborative project should never be taken lightly. A typical project involving say five partners, three industrial and two academic, to run for three years with a budget of $1million will require an experienced project manager devoting at least half of their time to the role. If the project is more complex or involves collaborations that cross national boundaries, such as a European Framework project, then the position of project manager will be a full time job. In addition to the heavy time requirements, the project manager also needs a certain degree of gravitas or clout that will allow them to control the project and deal with all the partners. The project manager needs to be taken seriously by the whole team. Therefore, although it is tempting to give the job to a junior member of staff, better results will come from the appointment of someone with experience and who commands the respect of all the consortium members.

The project manager is responsible to the steering committee for the day to day running of the project. The most visible aspect of this is reporting progress back to the steering committee and to any of the other stakeholders such as funding bodies. The ideal person will therefore be someone with an eye for detail, good interpersonal and organizational skills and who is capable of writing clear reports.

Communication

The secret to good collaborative project management is good communication between the partners. In fact monitoring officers often use the state of the communications within a project as a gauge to how well the project is working, since an early indicator of problems is often a breakdown of communications.

Regular technical meetings in addition to the main steering meetings are an excellent way of managing progress but this can sometimes be inefficient if partners are not local to each other. When face to face meetings are not practical, regular video conferencing or voice conference calling is a good way of ensuring regular conversations.

Electronic communications are also an excellent way of keeping everyone informed of progress.

In addition to a project website that might contain a 'partners only' area for posting progress information, many projects create a Wiki on which all the project members can quickly chat and leave messages and notes about issues within the project. Email groups and instant messaging are also common tools in addressing issues and keeping everyone up to date.

Keeping up to date milestone and deliverable status information and a copy of the risk register on a private website or Wiki is also an excellent way of keeping everyone aware of the pressing issues and deadlines within the project.

4.5 Working with monitoring officers

Monitoring officers are often appointed by funding bodies to ensure that grant funds are being spent appropriately and to support the project in their dealings with the funding body.

In the vast majority of cases, the monitoring of the project is supportive and positive. There is little point in a funding body supporting your project only to have a monitoring officer try to trip you up and find reasons to cut the funding. It is in everyone's interests for the project to succeed and the monitoring officer should be seen as a helpful addition to the team.

Monitoring officers are selected by the funding body for their experience and knowledge, not only of project management, but very often for their technical understanding of your project area as well. The chances are they have come across many of the problems you are likely to encounter and will be able to help you overcome them.

Monitoring officers also have their own objectives for your project and they are worth considering too. Most funding bodies use external monitoring staff. That is they are not employees of the funding body, they are consultants who have to compete for the opportunities available. Being a successful monitor therefore relies on working with successful projects and on positive feedback from the supported project about the added value and support provided by their monitor. It is therefore in the monitor's interests also that the project does well and that their contribution is appreciated.

What all this means is that you should keep the monitoring officer fully informed about the progress of the project and not be afraid to ask for advice and feedback as you need it.

We will look at the management of project problems in chapter 7, but it goes without saying that in virtually every scenario, the monitoring officer will play a key role in helping you through the problem. Keeping them up to date and honestly informed can only be of benefit.

4.6 So in conclusion

1. Sound project management is especially important in collaborative project. Of particular importance are:
- Project control
- Proper planning
- The management of risk
- Project leadership
- Working with monitoring officers

2. Project control is achieved through:
- A steering committee structure
- The appointment of an experienced project manager

3. Detailed planning should be carried out on each work package as it starts, the same planning techniques described in chapter 3 can be used for this 'rolling wave' planning.

4. Project risks consist of two types:
- Operational risks
- Insurable risks

5. The management of risk involves:
- The identification of risks
- Impact analysis
- Developing strategies for forecasting risk
- Developing strategies for mitigating risk

6. Project leadership is essential and requires:
- A competent project manager
- Good communication between the partners

7. Monitoring officers are a good source of:
- Experience
- Contacts
- Problem solving ability

5 Working with academics

A great many collaborative projects involve academic partners as the source of new technology or to exploit a specialist skill set, resource or facility. This chapter is provided to benefit industrialists who have not worked with academics before. It is not that academics are strange, it is just that they have different motivations and attitudes from industrial or commercial partners. Once understood however, they are formidable partners and a source of outstanding capability.

In order to convey this information, I am going to have to generalise a fair bit. I do not want to offend any academics in the process, I used to be one and have worked with a great many over the years. I hope you can use this information to recognise and understand certain traits so that you are firstly not surprised, and secondly can harness the capabilities that academics offer as effectively as possible.

5.1 Academic objectives

One of the first things that stand out about the objectives of academics for being involved with collaborative projects is that the majority of them are qualitative. That is they tend to focus on topics such as:
- Finding a useful application for their research,
- Providing opportunities for PhD and post-doctoral study,
- Improving their personal standing and the prestige of their institution,

- Finding a mechanism by which to further their research interests,
- Publishing application based papers on their work.

Where financial objectives are involved, they tend not to be related to profit or income as much as to fund PhD students or research staff.

This is not to say that academic institutions are not interested in seeing a financial return from their involvement in a project. Universities are becoming increasingly interested and demanding of a financial return. Many institutions have sophisticated contracts and technology transfer departments which work hard to secure a fair return on project results in the form of royalties.

So what does this mean for your relationship with an academic partner? Firstly it means that the individual academic staff you work with will not be motivated by the marketability of the results as much as the research process itself. This means that in order to get the most out of their involvement, you will need to keep them interested in the project work and technology, rather than the eventual results or their exploitation.

Given this initial lack of interest in the exploitation of the project, it is very important that the academic members of your team fully understand the wider objectives of the collaborative project and how their work fits into achieving them. The best results are achieved when this extends to the research staff and students in addition to the principal investigators and professors.

In at the deep end

Arrange for all the research staff to spend some time at the industrial partners' premises, getting as close as possible to the work the project is designed to assist. I don't mean give them a tour one afternoon, but actually have them spend a week or two immersed in the environment. The benefits this can reap for the project can be profound.

For example, years ago I ran a collaborative project to develop intelligent ultrasonic diagnostic technologies for the non-destructive testing of steel welded joints. In the first month of the three year project, we arranged for the PhD student working on the project to do a 2 week residential training course to become a certified ultrasonic testing practitioner. The understanding of the process and the knowledge of the issues that he came back with had a profound effect on the direction of the project and its eventual success.

Even if a baptism of this nature is not practical, a deep understanding of the problem and the formation of good working relationships can still be achieved with short term secondments or shadowing. Try placing the research staff in the industrial partners' development centre or shop floor for a week. Allow them to work on fire fighting projects or in quality control where they can see for themselves the issues that are being faced by their industrial and commercial team mates.

Publications

Publications are a very important part of academic life. Individuals and departments are often judged on their publication record. Your academic partners will therefore be very keen to write papers for conferences and journals about the work they are doing with you. Fortunately many understand the impact that publications can have on patent protection and commercial success. It is however very worthwhile to write terms into the consortium agreement that describe the publication process to allow industrial partners to suggest edits or even veto dissemination that could damage their prospects.

A brief discussion of up-coming publications should be a standard agenda item at steering committee meetings.

5.2 Attitude to risk

Most academics' tolerance to technical risk is very high. It has to be in order to continually push the boundaries of science and technology. In most cases, their tolerance to risk will be much greater than their industrial counterparts. Most research is paid for by grants and so there is very little financial risk involved in a project that fails to achieve its technical objectives. Most research funding bodies are happy to accept genuine technical failures as legitimate, allowing staff and institutions to continue receiving grants even after previous projects have failed.

It is important therefore that the industrial partners fully understand the level of technical risk involved in the research aspects of the project that the academics will carry out. This could be a matter of semantics, low risk to an academic may be unacceptably high to an industrialist committing their own scarce resources to the project.

Financial risk on the other hand, is a completely different story. Most academic institutions are incredibly risk averse when it comes to funding. A common manifestation of this can be seen in the refusal of academic institutions to even advertise a job position until all the offer letters and consortium agreements have been signed. This can cause severe delays at the start of a project as the process can take several months. Looking in from the outside, industrial partners often despair about the time this takes and the apparently avoidable delays that are caused. Understanding this limitation up front is therefore an important part of the planning process, be realistic about how quickly the project can start after the contracts are signed.

5.3 Agility

Industrial companies are often highly agile and capable of changing direction and focus to suit a new market opportunity. Academic partners may not be quite so willing to change direction so readily.

In many cases, the work will actually be carried out by Masters or PhD students. In these cases, the programme of study has been carefully worked out and approved as being of sufficient technical and academic merit to allow them to achieve their qualification. Any changes of direction that could downgrade the merit of the work will therefore cause concern to the student and their academic supervisors.

Using post-doctoral research staff does not entirely avoid this problem as they tend to want to work on projects that can lead to their first lectureship and launch their academic careers. If the project changes to such an extent that it no longer allows them to carry out valuable academic research or to produce high quality publications, they are likely to resist the change or even move away to another opportunity.

Changes that alter the scale of the work to be carried out also cause significant problems for academics. Research groups are funded by many projects and grants. Normally each individual student or research assistant will be funded by a specific project. If the demand from the project were to half, there is not normally any spare funding available with which to cover the remainder of their salary costs. Equally if demand were to increase, there is seldom any spare resource that could take on the work. Appointing new staff to carry out the additional work would involve the recruitment process once again, which could not start until the increase was contractually binding.

Changes to the duration of projects are also problematic. Most universities will not allow a PhD student to start work if there is insufficient contracted time available for them to complete their studies. Most academic employment contracts are therefore for long periods, typically 1 to 4 years and usually cover the entire planned duration of the project. If a project is shortened, the institution is unable to shorten their employment contract and will therefore be exposed to costs it cannot recoup. It also follows that practices such as probationary periods or the replacement of staff that don't work out as hoped are very difficult to implement.

The prospect of terminating a failing project will not be looked on favourably by academic partners. If they have issued long term employment contracts to their staff, they are likely to strongly resist early termination of the project. It is for this reason that industrial partners may require the consortium agreement to allow termination on the unanimous vote of the industrial partners only. Industrial partners would not want to be forced to continue to invest in a failed project just because the academic partner wants to cover their employment liabilities.

5.4 Academic seasons

The academic year is highly seasonal. The year starts in October when the new intake of students arrive and most academics are therefore extremely busy from mid September through to December. The period from January to April is a little easier although there is a lot of teaching going on. May to late June is the period when most examinations take place and therefore many academics are busy with marking and moderating activities. Come July and August and you hit the holiday season which runs into September. This time is popular for conferences and of course summer schools. Finding a good time to get academics' complete attention can therefore be a little difficult, but understanding these seasons is the first step to coping with them.

The fact that a large proportion of projects involve post-graduate and post-doctoral students also presents limitations on when projects can actually start. Most degree programmes finish in the early summer so candidates for research work are available for a limited period. Recruiting in the spring will be too early for the current cohort and the good ones from last year will already be employed. For this reason, academic projects tend to start in late summer or early autumn. It is not uncommon for collaborative projects to be forced to fit in with this timetable to suit the academic recruitment cycle.

One more observation about the use of post doctoral students is that unlike degree programmes, their PhD study is seldom run to a strict timetable. You may find that when the student starts working on your project, they are still writing up their thesis in their spare time. Activities such as final experiments and vivas could eat into your planned resources. This is standard practice and so needs to be factored into the planning of the project.

5.5 Ownership of IPR

Academic institutions are often very keen to retain the intellectual property rights to any inventions created within their projects. The reasons for this are that developments go on to be used in their teaching activities and as background IP for future research projects. This can sometimes cause issues with commercial collaborators who see the ownership of IPR as crucial for their security and in the case of small companies, to assist in the raising of capital.

When a license has been issued by an institution, it will expect a royalty payment to be made once the IP is exploited. This will still be the case even when the institution has had all its costs paid through a grant. One of the few exceptions to this rule would be if a company was paying an academic to carry out work as a subcontractor or consultant. In these cases it is usual for the company to obtain any IPR created. However, it is worth ensuring this is the case by writing it into the contract.

Academic institutions do sometimes agree to transfer or assign IPR to collaborating companies to assist in the exploitation but often require some form of insurance in case the company fails to exploit it. In these cases it is possible for the institution to grant an exclusive right to the IP initially and commit to assign the full rights once the company has achieved some reasonable milestone, such as the launch of a product.

5.6 Academic finance offices

In general, cash flow is not something that academic institutions worry about a great deal. It follows that there is seldom great pressure applied in making timely grant claims or financial expenditure statements. Now this is a generalisation and I have to say that in my experience, some institutions are highly professional and efficient. On the other hand, some are absolutely appalling and seem incapable of making an error free claim within a year of the work being carried out.

The risk is that grant payments to industrial partners, and other academic institutions for that matter, could be seriously delayed if you are unlucky with your academic partner's finance office.

It could be very worthwhile arranging a meeting between the project manager and the finance departments of the academic partners to make sure they are briefed on exactly what is expected from them and when. If there is an appointed monitoring officer attached to the project, they may help with this process as they may be able to define the claims procedure and rules with more authority than the project manager.

5.7 So in conclusion

1. The following aspects of working with academic partners should be understood:
- Their objectives are mainly qualitative
- Their tolerance of technical risk is high, but their tolerance of financial risk is very low
- Their agility is sometimes limited
- Their year is highly seasonal
- Ownership of IPR is important to them
- Academic finance offices differ in quality and may require briefing to avoid delays

2. Their objectives are likely to include:
- Finding applications for their research
- Providing opportunities for PhD study
- Personal and institution prestige
- Furthering research interests
- Publishing papers on the project

6 Working with industrialists

This chapter is provided to benefit academics who have not previously worked with industrial or commercial partners on a collaborative project. Again in order to convey the information I am going to have to generalise a little.

6.1 Commercial motivations

Being in business is about making money. Companies exist in order to provide their shareholders with a return on their investment, either in the form of regular dividend payments which distribute profits, or in the form of increases in the value of their shares. Company directors' first duty is to act in the best interests of the shareholders, in other words to ensure that the company makes them money.

It stands to reason therefore that companies will only get involved in collaborative R&D projects if there is a return to be made that will benefit their shareholders. This explains why the common and principle objectives of companies within R&D projects are predominantly quantitative.

The directors' duty to act in the interests of the shareholders extends to the running of the project after the initial commitments have been made. The company should continually review the relevance of the project to the direction of the company and ensure that the outcomes will still be of benefit. If the project starts to fail or loses its relevance, then it is the duty of the directors to either make changes to the direction of the project to keep it aligned with the direction of the company, or to cease investing in it. It is not uncommon for companies to pull out of projects for these reasons and we will look at how to deal with this in the next chapter.

6.2 Agility

In general companies find it much easier to change direction and respond to new opportunities than do academic partners. Most contracts of employment require only a one month notice period and there is seldom a problem switching individuals between different projects and activities. New staff are normally given a probationary period in which to prove themselves and secure longer term contracts. The ability of companies to quickly expand and contract to suit both their commercial and project needs is therefore greatly enhanced.

There is however a downside to this level of agility. Companies can be easily distracted. It is not uncommon to find that someone you thought was dedicated full time to the project, has actually been dragged off to work on something more pressing with little or no notice. This is particularly true in small companies where technical personnel are often called upon to fix product problems or help satisfy a large order.

Good communications are therefore very important to ensure that everyone is aware of the company's focus and the continued effort that everyone is able to provide. The project manager should keep themselves aware of who is working on the project to ensure that they receive all the relevant information and resources they need. New participants should be included in any meetings with other partners as soon as possible to introduce them to the rest of the team and to ensure they fully understand the objectives of the project.

6.3 Cash flow

In business, cash is king! If a company runs out of cash and is unable to pay its suppliers and creditors, it becomes insolvent and has to call in the receivers. A lot of effort and careful attention is therefore paid to the accounts and the flow of cash in and out of the company. Obviously the smaller the company the more susceptible they can be to the problem, but cash flow is important to all businesses.

Problems can occur with cash flow even in apparently highly successful and profitable companies. The principle drivers come not from the margins a company can make, nor their sales success, but in the time that elapses between them having to pay their suppliers and them receiving payment from their customers. Imagine a company that makes widgets for $5 and sells them for $10. If the customer pays up front and the raw material suppliers are paid later, then there is no problem at all. However if the suppliers need paying up front but the customers don't pay until 30 days after delivery, the company is left with a cash shortage between the two events. Sudden large orders or sustained sales growth can put huge demands on cash flow. Companies need to have an healthy cash balance or working capital in order to survive.

Grant payments are normally made in arrears and typically only once per quarter. This means that the company will be spending its own money on salaries, materials, even large capital equipment purchases, often up to six months before they will receive their grant contribution. This has two important ramifications. Firstly, each company in the consortium needs to be cash rich enough to be able to afford to take part in the project. Secondly, the process of claiming for the grant each quarter needs to run like clockwork if un-due strain is not to be exerted on the commercial partners.

> **Getting touch on claims**
>
> In many cases, grant claims are made by a lead partner who, once paid, distributes the grant to the others. If the claim is delayed due to one partner being slow with their claim data, everyone has to wait.
>
> Put pressure on the partners by setting a deadline by which all claim data is required. Those that are late miss the boat and have to wait until the next claim period. This will allow the efficient partners to be paid without delay and focus the attention of the slower ones.
>
> To operate a scheme like this, you may need to write the details into the terms of your consortium agreement. That way all the partners are fully aware of it and have given their approval.

The commercial attitude to cash flow is in direct contradiction to the attitude of academic institutions. It is therefore very important that all partners within your consortium understand the pressures and work together to ensure that grant claims are made in a timely manner.

6.4 Micro companies

Micro companies include very small start-ups with only a handful of employees, specialist contracting groups and individual self employed consultants. It is not uncommon to have such small enterprises involved in R&D projects because they tend to offer exciting new technologies or specialist skills not easily found elsewhere.

Micro company involvement in collaborative projects can be very valuable, however they are even more sensitive to the agility and cash flow issues mentioned above than normal.

Technology based start-up companies tend not to have revenue streams from product or service sales. They might have venture capital investment or are being supported by their directors own funds. It is only fair that the other members of the consortium satisfy themselves that they have the financial strength to be involved in the project and that they will be able to deliver against their promises. This might extend to their ability to exploit the results as well as simply carrying out the project work.

To be involved in the project, the outcomes must be 100% aligned to a start-up company's business plans. A start-up that hops from one exciting project to the next will not stay in business for long. The other partners may therefore wish to understand the motivations and objectives of very small companies before committing to work with them.

Small consulting groups and individuals can also be a cause of concern. Whilst they often bring a specialist knowledge or skill set to the project, you will need to be sure that they can cope with the cash flow issues mentioned above and that they have sufficient additional income or resources to be able to provide their share of the investment in the project.

It is not uncommon to see consultants claiming that their time is their contribution. For example they may say that their rate is $500 a day and that they will put in 100 days of effort, thus make a contribution of $50,000. Grant funding bodies often take exception to this kind of arrangement, especially when they are expected to pay grant against it. The concern is that the consultant is not paying themselves $500 per day but is actually surviving on the grant alone. Effectively the consultant is not investing anything in the project themselves except a nebulous 'loss of potential income'. Funding bodies are therefore particularly cautious of new consultants with no trading history or other visible means of support.

The same issues can sometimes be found in very small companies where the directors and staff are paid very small salaries but given generous share options or dividends, often for tax efficiency reasons. Funding bodies do not often accept shares and dividends as a direct salary cost as strictly speaking they are a profit distribution and therefore ineligible as a project expense. This can mean that the grant is severely reduced. For example imagine a director of a small company who takes a salary of $5,000 but is then paid $45,000 in dividends. They might claim that their salary is based on the total $50,000 resulting in a day rate (from table 1.1) of $258. If they do 100 days work that would be a cost of $25,800 of which they could get a 50% grant worth $12,900. The funding body would take a different view saying that the eligible salary was just $5,000, the day rate $25.80 and therefore the grant would be just $1,290. This drop in grant is likely to out way any tax advantages.

Partner or sub-contractor?

There are potential solutions to the problems associated with involving micro-companies in your project. Using a consultant or start-up as a sub-contractor rather than a partner is one of them. As a sub-contractor they could receive full payment for their involvement in the project. Of course there has to be a trade off and this often comes in the form of less beneficial access to the project results or new IPR. There is potential for negotiation however, between access to the project results and the rates paid for the work. For example if the sub-contractor carried out the work at, or even below cost, they could have access to more of the benefits than if they charged full commercial rates.

It might also be possible to mix the role of partner and sub-contractor. They could be fully paid for some aspects of their work but act as cost sharing partners for other aspects. Their access to be project benefits would then be limited to the cost sharing aspects.

6.5 So in conclusion

1. The following aspects of working with Industrial partners should be understood:
- Their objectives are mainly quantitative
- Their agility can be very high
- Cash flow is very important

2. Micro companies present special issues for collaborative projects including:
- Limited or no revenue streams
- Ease of distraction
- Commercial vulnerability
- Exploitation strength

7 Managing problems

Projects always carry risks. R&D projects are particularly risky due to the fact that the innovations might not work as hoped. In addition to this fundamental risk, collaborative projects also suffer from their own unique set of issues and problems. This chapter covers some of the common pitfalls and provides guidance on how to spot, manage and hopefully overcome them.

The first thing to recognise is that problems are common place and you will need to be vigilant to spot them and deal with them in time.

7.1 Partner problems

It may come as a surprise, but one of the most common problems encountered by collaborative projects is the withdrawal of one of the partners. Even more surprisingly, one of the most dangerous times for partner loss is before the project has even started.

A common cause of partners failing to start a project is that the consortium cannot agree the terms of the consortium agreement. Areas such as the ownership and access to the new intellectual property and the share of liabilities are often areas of contention between partners. It is therefore very important to start discussions about the consortium agreement as early as possible and certainly before any grant applications are made.

When to bring in the lawyers?

The consortium agreement is a legal document that binds the partners and therefore all partners must seek legal advice in its drafting and agreement. However, the point at which the lawyers get involved should be carefully managed. Too early and they can create arguments and bad feelings within your nascent consortium before any work has even started.

In my experience it is best for the partners themselves to discuss and agree the basic terms of the agreement first, then bring in the lawyers to paper it up properly and check for any problems.

Another cause of failure at an early stage is lack of internal company communications. Sometimes the technical staff can get very excited about a project and pursue the negotiations and planning without getting the full approval of their senior management or board. When the project is finally brought to the attention of the senior management, it can come as a severe shock when they decide to pass on the opportunity. Quite obviously, good communications can help avoid this problem. Make sure both you and your partners are keeping all the key purse-holders informed with what you are planning, why you are planning it and how much it is going to cost.

This applies to academic partners too. University research offices very often have to sign off on costs and contracts. An individual academic may say they can carry out work for one price, only to find that the research office insists on full overhead recovery and the costs are actually quite different. Make sure these offices are involved early to avoid any confusion.

Once the project is underway, partner problems can unfortunately still occur. There are many potential reasons for partners to lose interest in the project. Some of the most common include:

- Changes in company structure or the movement of personnel
- Merger and acquisition activity
- Changes in business direction or motivation
- The company falling on hard times

When companies go through structural changes, the rational and commitment to carry out existing projects often needs to be re-established. This is particularly true when the responsibility for the project moves from one department or division to another. The situation is made worse still if the re-structuring necessitates changes in the personnel working on the project, particularly the project's champion within the business. This can be a very dangerous time for a partner's role in your project.

One way of coping with the change is to recognise as early as possible that it is happening and to pro-actively work with your contacts in the organisation to make the transition as smooth as possible. The existing champion will firstly need to reaffirm the objectives of the partner for being in the project and confirm with senior management that they remain valid and important. If a change in personnel is involved, the existing champion, together with the project manager, should meet the intended replacement and 'sell' them on the project, its progress to date and the objectives. The more the new people understand about the project and its benefits, the better the chances of a smooth transition.

This is however likely to be a very busy, if not stressful time for both the existing and new personnel. Finding time to talk about legacy projects may not be easy, so the project manager has a role in ensuring it happens and helping to facilitate the hand over. It is after all, important that the project manager quickly forms a good working relationship with the new personnel.

Similar pressures occur when a partner is taken over or merges with another organisation. After an acquisition, the new owners tend to make sweeping changes throughout the organisation and often this affects legacy projects that the company is involved in. The process to deal with this is very similar to the process described above for re-structuring. The internal project champion needs to first re-affirm the project objectives and try to structure them in a way that will be attractive to the new owners. Once achieved, the champion and project manager should meet with the new management and 'sell' the project to them to gain their commitment. Some changes to the focus of the project may be required to suit the new owner's new objectives for the company.

Changes in business direction are much harder to cope with. In this scenario the company shifts its focus in such a way that their objectives for being in the project can diminished or disappear entirely. This could be caused by all manner of commercial or technical reasons such as:

- the emergence of a cheaper alternative technology,
- the need to focus resources on a more pressing problem,
- a change in the market dynamics that adversely affects the commercial opportunities of the project,
- the emergence of a more exciting opportunity elsewhere…

If this occurs, the only chance for their continued involvement is to change the project in such a way as to make it attractive again. This is more easily said than done however, as the other partners may not wish to see the project change too much. In these cases the project manager needs to work closely with all the partners to try to refocus the project to everyone's satisfaction. A steering committee meeting should be called as soon as possible to discuss the issue and agree any changes in the project's direction.

If a partner company falls on hard times, perhaps through declining sales or other failures, they may feel the need to cut back on activities that are not generating immediate income or profit. R&D projects are often high on the list of 'non-essential' activities that can be cut back on. In this situation it may be possible for the partner to remain in the project but reduce their commitment to it, either temporarily or permanently. Call a steering committee meeting to discuss how best to move forward.

Know when to call it a day

Up to a point it is best in all these circumstances to try to resolve partner issues by selling the project to new management, changing the focus of the project to suit changing situations, or adjusting the level of involvement for particular partners. However, there comes a point when the changes required in order to keep a partner interested in the project become too much for the rest of the consortium to bear.

Forcing a partner to stay with a project because they signed an agreement is likely to be highly counterproductive. Without the proper motivation, their contribution is unlikely to be helpful or productive to the project. Morale within the other partners will also suffer as a result.

If there is no reasonable way to keep a partner involved in the project, the rest of you should let them go. The consortium agreement should allow this to take place and describe the process. You should then work together to either re-distribute the work or find a replacement partner.

If a partner does decide to pull out of the project, the project manager should call an emergency project steering meeting as soon as possible. The implications to the project should be discussed and understood and a recovery plan put into effect.

The first thing to look at is the ramifications to the existing project plan. All the tasks that the departing partner was going to be involved in will need to be identified and examined in turn to decide:

- if the task is still relevant, it may be that it only satisfied the objectives of the departing partner, in which case it could be deleted from the plan,
- if it is still relevant, how will the task be affected,
- and what work could be re-distributed to the other partners,

If it is possible to completely redistribute the work between the remaining consortium members without adversely affecting the project outcomes, and provided the partners have the resources, then this is the best approach to take. If however the partners cannot fill in all the gaps then an alternative will need to be found.

If the project is grant funded it is important that you notify the funding organisation and any monitoring officer at this point, firstly to keep them informed, and secondly to ask for their assistance in looking for a solution.

Providing relations with the departing partner are not too strained by this stage, one possibility might be to ask them if they would be prepared to do the work as sub-contractors. Obviously they could not share in the benefits or any new intellectual property but they might be willing to at least carry out their remaining work for a reasonable fee.

If this is not possible, then a new partner will need to be found and brought into the project. It may be necessary to halt the project until a replacement partner can be found. This may cause difficulty for partners with dedicated project staff that cannot be temporarily re-deployed and so will need to discussed and agreed by the steering committee.

Finding replacement partners

The process of finding a replacement partner is rather like repeating the process of building the consortium in the first place. It is best to involve all the remaining partners in generating a list of potentials and seeing if anybody has any contacts that could be useful in making an approach. If the project is grant funded you should involve the monitoring officer too, they may have good contacts that might be helpful.

The approach to a new potential partner will be a little different from normal as the work is already underway and you are looking for someone to get involved as quickly as possible. On your side is the fact that the project is already running and that a new partner could therefore benefit from the work that has already taken place. There may also be guaranteed grant contributions available.

You will need to have a convincing argument ready to explain why the old partner left the project. The new partner will want to reassure themselves that they are not making a mistake by jumping into a project that others have managed to escape. You may even be able to arrange for the old partner to place a call explaining their reasons, just to put their minds at rest.

Since the project is already underway, confidentiality may be very important. You should have potential partners sign non-disclosure agreements before you tell them too much about the project and your early results.

When the new partner joins, the project plans will usually require some modification to suit the resources and motivations of the new partner. It is best to involve all the partners in this process so that everybody can understand and approve the necessary changes. Hold a full steering committee meeting inviting the new partner to meet the whole team and get involved straight away. Treat the meeting as a celebration of their involvement and spend plenty of time describing the work carried out to date to bring them up to speed. Describe the aspects of the project plan that they are to be involved in then run a re-planning workshop to modify and update the project plan.

Fortunately, academic partners do not often pull out of collaborative projects. Their motivations and focus are not easily shifted and even re-structuring is unlikely to affect a running project. There are however occasions when a key academic might move institutions in the middle of a project. If this happens there are two basic alternatives. A different academic in the same institution might be able to take on the project. If this is the case, the project manager should meet with them to ensure they are both competent and committed to the project.

The alternative is for the project to move with the academic. In other words, for the academic's new institution to replace the old one in the consortium. This is more complex and would rely on the old institution being happy to lose their interest in the project and any potential intellectual property that might be generated. The research support offices of both institutions will need to be involved in this negotiation. Again the project manager needs to get involved to help negotiate a transition that is beneficial to the smooth progress of the project.

Just one further thought on partner problems. Sometimes it is the project's lead partner that is the one to lose interest in the project. To be fair to the other partners, as soon as a change is recognised, a steering committee meeting should be called and the roles of project lead and project manager transferred to one of the other partners if necessary.

7.2 Relationship breakdowns

It is a sad fact of life that occasionally partners can fall out. This can happen at an individual level where there is a clash of personalities or misunderstandings or at the corporate level.

Corporate level disagreements could come from anywhere and be about anything. They might range from:

- the breakdown of another project that spills over into yours,
- legal action being taken between them
- competition in the marketplace,
- a negative advertising campaign,
- a problem over common standards adoption.

The project manager should be on the lookout for signs of such problems. If the conflict is at a senior management or board level there will be little that the project can do to calm things down, except it might be useful to remind the parties of the good work they are doing together.

The poaching of staff can also cause conflict between partners. In this scenario one partner tries to recruit the key technical staff of another. This can cause such severe resentment and problems that some projects build a non-solicitation clause into their consortium agreement to prohibit such actions.

Conflicts between individuals in the project can be just as damaging but fortunately are easier to deal with. Again the project manager should try to identify when problems may be occurring, sit the parties down and try to broker a truce.

Identifying conflicts

The first signs that a project may be suffering from a breakdown in relations are the level and spirit of communication between the partners. If the partners are regularly communicating and all contributing to group discussion boards, then things are probably going well. If one or more partners appear to be distant from the rest or you start to hear comments such as; 'Who knows what they're up to'; 'I've been waiting for the widget for months, but you know what they're like'; or 'I tell you what, I'm not working with them again'; then things are probably not going so well.

As a project manager, talk to each party and find out what the situation is. Try to broker a compromise and bring the parties back together again.

Poor or devious reporting of progress is also a very good indicator that something is not quite right. If you have difficulty getting a response from a partner or if their reports of progress are vague, push them for an answer or even visit them to find out if there is a problem. Once identified, help them to resolve it.

Project meetings should always involve lunch. This provides the project manager and any monitoring officer an opportunity to judge the spirit of the collaboration and spot any issues that may be brewing between the partners.

7.3 Delays and overspends

Delays often cause stress to the relationships between partners, especially where one partner is waiting to start their activity and can't because one of the dependent tasks is not completed on time. Again good and regular communication of progress will give everyone an insight into when milestones will be reached and tasks completed. Very often if the others are aware of a delay, they will pitch in to help bring the task back on track or be able to schedule their plans to accommodate the delay.

At the beginning of the project, you create the project plan that you intend to follow. The plan is your best guess at the time as to how long things will take and how much they will cost. When you made the plan you may have built in some contingency and erred on the side of caution a bit, but inevitably some things will take longer and cost more than you thought.

It is very important to track time and cost as the project progresses. This will enable you to gauge whether your initial estimates were accurate and what implications there are likely to be if they were not. The mechanism for this is the regular monthly or quarterly progress reporting that was described in chapter 4. Ask all the partners to declare the resources they have committed and compare them to their initial plan. If they tended to underestimate, ask them to revise their estimates for future tasks and assess the impact on the rest of the plan.

As already mentioned it is very important to keep everyone informed of progress, especially the partners who will be working on dependent tasks. If they know in plenty of time that they may be delayed, they will more easily be able to re-schedule the resources or switch the order of their other tasks to accommodate the delay.

Re-planning

Re-planning is a common occurrence in many projects and is required after delays, partner changes or simply to make use of the best current information.

If you are employing the rolling wave detailed planning approach, then project level re-planning may be necessary as the latest estimates of future work package durations are produced.

The re-planning process is no different from the original planning process except that the basic tasks and milestones are usually still appropriate. It is worthwhile though going back over them all to ensure that the early assumptions are still valid and to add any additional tasks that may have become apparent. When thinking about each one, ask questions like 'given our progress to date, are these resources and times realistic?' and 'how can we better control the future work so that we don't get into the same trouble again?'

When delays do occur, it is important to bring the team together to adjust the project plan in order to recover the situation and if necessary agree on an extension to the overall length of the project. This will have implications for the costs and spend profiles for each partner. Updated plans should be agreed unanimously to maintain the common commitment of all the partners.

7.4 Technical failures

All research and development projects are risky because they contain elements of innovation that stretch the current state of the art. It is unfortunately inevitable therefore, that from time to time, tasks will fail because it is not technically possible to solve the problem being addressed in the way anticipated.

Such failures are known as legitimate failures because they are nobody's fault. The steering committee must meet to discuss what should happen next. The various alternatives might include:
- employing a different technology to by-pass the problem,
- changing the objectives of the project to reduce the scope,
- developing the partial results achieved to date,
- or winding the project up.

Any funding body and monitoring officer should be informed immediately.

7.5 Market changes

The original raison d'être for the project was based on a commercial opportunity described in the project's business case. If the market opportunity that was described changes during the life of the project, the project must either adapt or its termination should be considered.

It stands to reason therefore that the business case development and exploitation planning should be carried out as a parallel exercise to the technical work rather than leaving it until the end of the project. When problems are detected, they must be brought to the attention of the steering committee to decide what action is required.

The steering committee may decide that reporting on the evolving market understanding and exploitation plans should be a regular agenda item in order to maintain focus on this critical aspect of the project work.

7.6 How to avoid problems

It's a bit of an old cliché, but the best way to get out of trouble is not to get into it in the first place. Within collaborative R&D projects, a lot of trouble can be avoided through strong project management and good communications.

I hope this chapter has highlighted further the importance of the role of the project manager. It is a serious one that should only be taken on by an individual with the experience and charisma to be able to pull it off. The project manager needs to command the respect of the partners, be able to broker compromises when partners disagree, be able to spot problems and deal with them rapidly and maintain a good spirit within the project at all times. The project manager should also not fear calling a steering committee meeting should they feel things need to be discussed. Above all the project manager needs to be allocated sufficient time to be able to carry out the role.

A well defined project plan is also a key aspect of strong project management. If the plan is too loose or the deliverables are poorly defined, there is a greater chance of misunderstanding and failures. The more effort put into the preparation, the more smoothly the project is likely to run.

If the project has a monitoring officer appointed, this individual is likely to have a wealth of experience and can act as a good source of support for the project manager in both spotting and dealing with problems. A good monitoring officer can also act as an 'honest broker' in dealing with conflicts between partners. It is therefore important that your reporting to them is open, honest and timely.

Good communications are essential to successful collaborative projects. Newsgroups, Wikis, regular meetings and what ever other mechanisms you employ will have a significant impact on your chances of delivering the project on time, on scope and to budget.

7.7 So in conclusion

1. Problems can occur within projects for many reasons including:
   - Partners leaving the consortium
   - Relationship problems between partners
   - Delays and overspends
   - Technical failures
   - Market changes

2. Partners can leave the consortium before the project has even started due to:
   - Disagreement over the consortium agreement terms
   - Problems with the grant offer letter terms
   - Poor internal communications

3. Partners leaving a project mid way can be due to:
   - Changes in company structure
   - Movement of key personnel
   - Changes in business direction or motivation
   - The company falling on hard times

4. When a partner leaves the project, each affected work package needs to analysed in order to determine:
- If the task is still relevant
- How it will be affected
- What work can be re-distributed
- Is a replacement partner required

5. Relations between the partners can breakdown through:
- Corporate level disagreement or conflict
- Personality clashes between individuals
- Delays or over-spends that are not communicated adequately

6. Technical failures are often legitimate but actions could include:
- Employing a different technology
- Redefining the scope of the project
- Development of the partial results
- Winding the project up

7. Strategies to avoid problems include:
- Sound project management techniques and reporting
- Good communication between partners
- Appointing a strong project manager
- Working with the monitoring officer

8 Exploiting the results

Although this is the penultimate chapter in the book, exploitation is something that you need to be thinking about right from the start. There is very little point carrying out the project unless it brings you some form of benefit. Exploitation is about how you achieve and capitalise on that benefit.

The exploitation methodology described here is based on three distinct stages that start during the project planning phase and continue throughout and beyond the project. Before we can get into the exploitation methodology however, we need to understand what forms exploitation can take.

8.1 What counts as exploitation?

The precise nature of the exploitation of your project is likely to vary considerably between you and your partners. You each have different motives for carrying out the project and different objectives that you want to achieve.

The most common forms of exploitation are:
- Designing a new product
- Developing a new service
- Entering a new market
- Implementing a new process
- Licensing technology to third parties
- Publishing the results
- Creating opportunities for further development work
- Winning further grants

Developing a new product or service using the results of the project is a wholly appropriate exploitation route and one that can yield significant returns on investment. Equally exploiting project results through the expansion of an existing product or service into a new market can be highly beneficial.

Many projects develop techniques and processes that improve production and service delivery processes. Projects that result in reduced waste and emissions are also becoming increasing attractive as awareness of the causes of climate change improves. Here the exploitation is centred on greater efficiency, improved capacity or higher quality which can all lead to better margins or the ability to compete more aggressively in the market.

Licensing opportunities occur when the results of the project are not exclusively useful to the partners. Where the results could be useful to other companies and their activities would not create competition with the original partners you should consider licensing the technology. Normally, the results can be licensed in return for a royalty stream. Benefits of this kind are sometimes referred to as 'spill-over benefits'.

Licensing also occurs within the original consortium to allow individual partner's IPR to be used by the others in order to exploit their results. Academic institutions in particular are increasingly keen to see a royalty stream generated from their collaborative projects through licensing their results to their partners.

Publishing the results of research projects is an important aspect of exploitation for academic partners and some industrial partners including technology companies, research and technology organ-isations (RTOs) and charities. Publishing results helps to add credibility to the organisation and improves their standing in the community.

Individual academics and their departments are often appraised on their publication record for promotion decisions and budget allocations.

If your business is research based, then using the project results to feed future research is an important deliverable. It helps to sustain your business and research department. Again this is most common in the case of academic and research organisations, although many large corporate research laboratories also site this as an important deliverable.

Exploiting the results through the winning of further grant funding is often the case when the original project was highly pre-competitive or far from market. The original project may have proved a concept and provided the credibility and confidence to apply for larger grants to take the project to the next stage in its evolution.

8.2 When and what to plan

The exploitation methodology contains three distinct activities that run from the initial planning stage right through to the point of achieving the objectives of the project. The three activities are:
- the business case
- developing market understanding
- the exploitation plan

Figure 8.1 shows the three activities overlapping over the course of the project.

8.3 Business case

The business case was introduced in Chapter 3 because the majority of the activity takes place before the project actually starts. Building the business case is an essential first step because the findings will heavily influence the direction that the project takes and largely determines the list of objectives and required deliverables.

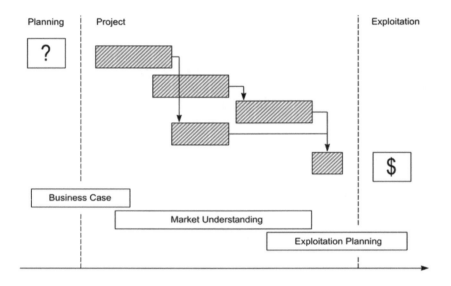

Fig. 8.1. Exploitation methodology

A key part of the business case is the analysis of the market opportunity. The analysis covers who the potential customers are, the problems that they currently experience and which the project will overcome, and the potential scale of the market opportunity. All of which are built on during the project.

8.4 Market understanding

While the project is running, it is essential to continue the early market analysis and understanding work in parallel to the technical activity. The reasons for this are very clear. You and your partners are investing valuable time and resources in the project in order to achieve a business benefit. The earlier you discover what the market wants from you the easier it is to focus the project to best suit your objectives. Even seemingly trivial decisions, like the choice of scenarios for an experiment or demonstrator, can influence the exploitability. If it makes no technical difference, let the market understanding decide who to target the demonstrators towards. Project monitoring officers often struggle to encourage projects to carry out market understanding and exploitation planning early in the project, only to hear later on: 'if only we had thought of this a year ago, we could have...'

The market understanding stage comprises of three parts:
- analysing the market opportunity
- understanding the market dynamics
- understanding the industrial forces

Analysing the market opportunity

The market analysis is the more complex to carry out, but it is arguably the most important of the three parts. Without customers you will never achieve your commercial objectives. Don't worry though because you made a good start on this analysis during the business case work. The analysis can be carried out through asking a serious of questions.

The first set of questions addresses the market need for your developments. This is a detailed look at who your customers are and what they want from you:

What problem are you going to solve?

As already discussed, whatever project you are working on, the exploitation will rely on your customers giving you their money in return for some benefit that you are going to provide for them. To make them do this, you are going to have to satisfy a need for them and one of the best ways of understanding this need is by describing the problem you are going to solve for them. An excellent way to achieve this is to build a set of 'use cases' for the target customers that you identified at the Business Case stage.

Use cases

Use cases are small documents that describe in detail an intended customer, their needs for the benefits your project will provide and exactly how they will deploy your solution.

The customers or end users may be represented within the consortium or advisory group, in which case it should be possible to describe in considerable detail the problems they face and the requirements that must be met. If this is the case, consider seconding research staff into the customer organisation for a short while so that they can really understand the issues.

If the intended customers are external to the consortium, you will have to rely on market research and empathy to understand their needs.

The use case should include short statements under each of the following headings:
- Description of the Customer – who are they, what market and industry sectors do they operate in.
- What problem do they have – describe in reasonable detail the problem they face that you hope to solve.
- How big an issue is it – how serious is the problem you are trying to solve and how often does it occur. What alternative solutions could be employed? How are they coping with the problem today?
- How will their behaviour change – describe how the customers will do things differently once your intended solution is in use. Take care to describe both positive and negative changes, for example are their cost, responsibility or standards implications?
- Stakeholder analysis – who and what is involved in the buying decision?

How will the customer's behaviour change?

This section of the use case is very important and needs to be explained carefully because changing customer's behaviour is normally a bad thing.

What we are looking for here is an understanding of the impact that your project outcomes are going to have on your customers. The wider the impact, the more changes are required to accommodate it, the more resistance there will be to accept it. Put simply – no one likes change. The more changes your solution requires on the part of the customers, the bigger the benefits have to be to outweigh them.

Take the example of television. When colour television was introduced, the designers were very careful to introduce a technology that did not disrupt existing customers with black and white receivers and provided a smooth transition as people bought into the new technology. They offered a system for new colour receivers with virtually no change in user behaviour at all. The new receivers took up the same space in the home, used the same power supply, the same aerial and gave access to the same programming. The customers adopted the new technology without much fuss, there was a clear benefit with minimal behaviour change.

Satellite television was a different proposition, to benefit from this innovation you needed an extra box, more remote controls, a different (and ugly) aerial, you even had to pay for the channels in a different way. Initially take up was slow and even now the industry has to offer more and more channels and services to attract new customers. The benefits have had to be large to overcome the disruption and costs. How difficult do you think it will be to convince us all to buy high definition television services and what benefits will they have to offer to convince us to buy? Ironically they are reverting to the original model, by providing high definition television through existing satellite and cable delivery systems. The disruption for existing satellite customers can once again be minimised.

So, what changes in behaviour will your customers need to make in order to make use of your project outcomes? Within each of your use cases, create a list and ask: Do the benefits make all this worthwhile? It is in asking questions like this that makes the creation of an advisory group invaluable. They should be able to tell you exactly how your solutions will change their behaviour and whether they feel it will be worthwhile.

Next, you need to compare the benefits to the list of behaviour changes in each case. Do they occur in the same place? For example if your solution is sold to an organisation, do the benefits accrue to the same individual or department as will need to make the necessary changes to accommodate it? If not, you could be in big trouble. To find out we need to look at all the stakeholders in that purchasing decision.

Who else is affected?

A stakeholder in this context is anyone that is affected by the purchase and use of your solution. Let's take the example of a new telemedicine device that will allow patients to monitor their own condition and regulate their own treatment at home. The immediate stakeholders are going to be the patients, the disease specialist and the family doctor. At first glance this can only be a good idea, the patient doesn't need to visit the specialist so often and the specialist and family doctor are able track their patients' condition more closely and receive early warnings of potential problems. But if we look at the changes in behaviour a little more carefully we find that all three will need to make significant changes to the way they work and live. The patient is taking on more responsibility for their own management, including the operation of test equipment. The specialist is relying more on remote data and less on physical examinations to make treatment and care decisions. The family doctor is also more involved than previously and takes on a responsibility to view and act on some of the data being generated. Will the benefits of providing better patient care outweigh these changes in behaviour and responsibility?

Perhaps using the telemedicine devices also requires some attention from the district or community nurse, will they need training, and do they have time? Is there an additional administrative burden on the healthcare provider? What are the IT implications? What do the healthcare authorities think, will they divert funding from the hospital to the family practice to pay for the equipment or its running costs? Before we know where we are, we have a growing list of stakeholders that includes the patient, the family doctor, the nurse, the administrator, the healthcare authority, the hospital consultant, the hospital administrators, patient support groups, charities, efficacy bodies even the insurance companies. Imagine them all sitting round a table discussing whether to buy the service for the patients. Who will be for it and who will be against it, and why?

This is a complex example, but it does illustrate the point that very often a buying decision is a complex matter, so how can you analyse it all properly? First draw out a diagram, like the one shown in figure 8.2, showing every link in the chain between you and the various stakeholders. For each one, write down what you think their reaction to the purchase will be and why. Try to think how you could convince them that it is a good thing for them that the purchase takes place. Are any of them likely to have a veto or raise an argument that will stop the purchase? If so their concerns become your highest priority. It may be that in order to satisfy everybody you have to change or expand the deliverables from the project. This will be very worthwhile if you can achieve a situation where as many stakeholders as possible are supportive of the buying decision and the remainder are neutral.

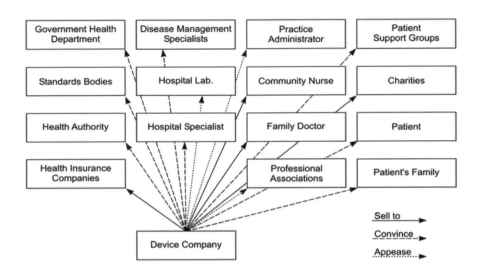

Fig 8.2. Stakeholder analysis

As always it is best to test your assumptions on real people. In the example above, it would be very worthwhile having repressentatives from the main boxes on the advisory group. Meetings should be arranged with representatives from the other groups too. Remember you can never understand too much about your customers and the problem you are trying to solve for them. Collecting the views of all your stakeholders is a very important part of this process.

Stakeholder analyses may need to be carried out independently for different territories. In our healthcare example, the analysis is very different for the US market than it is for any of the European markets, each of which differ significantly from each other too.

At this point it is worth taking stock and asking if you still believe that the project is relevant to this use case. To recap, you need to be convinced that the problem you are going to solve for your customers is real and serious, serious enough for them to spend money on your solution. You need to believe that your project outcomes will bring them benefit and value for money. Finally you need to be sure that the disruption your solution will cause will be acceptable to all the stakeholders that will be affected because of the larger benefits that will accrue. If not you should consider making changes to the course of the project, the scope of the deliverables or looking for alternative customer or end user groups.

Quantifying the market opportunity

The next set of questions addresses the market potential for your idea. Essentially we are going to get a better feel for the numbers you first created during the business case work. We will estimate how big a market you have and how much money you can make from it.

Where are your customers and how many are there?

So far, we understand your aspirations for the project, we have a list of your potential customer groups and we know what you are going to have to do to satisfy their demand for a solution to their problem. Now we need to know how many of them there are and where they can be found. At this stage we are looking for the total potential market size.

Depending on the scope of your project and the reach of the collaboration, you will need to find out these numbers for the region, country, trading block or even globally. If you are going to develop a product with potential global demand, you will need to know the market sizes in each of the major territories that could be of interest to you and your partners.

For each of the customer groups you have identified, find out how many there are within your range. If we use the earlier telemedicine example, you will need to know how many patients would be suitable for the product, and how many family doctors' surgeries there are in the country. You would also need to know how many hospital departments we will be affecting and how many health authorities there are.

Most of the information can be gleaned from the internet. Information on the general public can be found on local government websites. In the UK, the Office of National Statistics (ONS) website has a wealth of demographic and socio-economic data. Various other government publications and websites contain useful information, especially the Department of Trade and Industry and in the case of our medical example, the Department of Health and National Health Services websites. Trade associations often publish detailed statistics and reports on the state of their sectors. Professional institutions such as the Institute of Marketing and Institute of Directors have excellent libraries that contain up to date reports and market data.

Help can also be sought from local and regional government services such as regional development agencies that offer advice, local market intelligence and maintain lists of services and specialists. Some even offer grant support to commission market research work.

How is the market size changing?

When you are looking for market size data, try to get as much information about how the market has changed over the last 5 years or so and if possible collect predictions about how it will change in the future. This market dynamic information will be a very useful indicator as to the health of the market.

Market dynamics are important because they tell you whether there will still be an opportunity waiting by the time you have finished your project. Ideally you only want to exploit your results in markets that are growing or are large and stable. Declining markets should be avoided all together as they are unlikely to be able to provide you with a suitable return on your investment, or any long term potential.

What trends are likely to affect your business?

The world is not a static place and many of the changing trends that exist may affect your project, some of them positively.

One of the biggest economic changes affecting us over the next few years will be the aging population. The proportion of the population that are retired is the largest it has ever been and the trend is set to increase dramatically over the next decade. If your project is related to this group then your market size is going to expand dramatically. This change will have implications for healthcare, entertainment, housing and many other sectors. How will it affect your project objectives? Collect data from the Office of National Statistics to help predict the scale of the change to the market data you have collected already.

There are other trends such as climate change which will have an effect on many sectors. Milder winters and dryer summers will affect energy use, sports and recreation, agriculture and retail demand. More aggressive weather systems will increase demand for disaster recovery and prevention technologies.

Technological developments also act as trends, the proliferation of mobile data communications, universal broadband internet access, and the reduced size and power requirements of electronic devices are trends that may well have an effect on your project outcomes.

Regulatory changes have dramatic effects on markets. The introduction of requirements for waste management and sustainability has driven a whole new industry in mobile telephone recycling. What is the regulatory environment like for your sector and how is it likely to change in the future?

Let's take stock again. In order to proceed with this use case, you now need to be convinced that there are enough potential customers within your reach to achieve your ambitions, that the market is growing sufficiently to provide a suitable return on your investment and that any trends that you have identified will support the market and not threaten your hard work. If the answers to this point are positive, then that is very good news. It is now time to take a look at the Industry forces you may be working in.

Understanding the industry forces

In many if not most cases, the project outcomes will be exploited within the partner organisations' existing industry sectors. There will be some cases however where the project will lead to the potential to penetrate new industry sectors. In these cases it is important that an understanding of how the new industry sector works is developed.

The terms industry sector and market sector are often interchanged. However they are quite different. Industries consist of sellers and markets consist of customers. So far we have concentrated on analysing the market opportunities created by the customers.

What we need to look at next is the selling environment. This includes all the aspects of doing business that affect your ability to deliver your project results to your customers. Over thirty years ago, Michael Porter identified five forces that affect the profitability of an industry and the basic principles still hold true in today's economies. The five forces are:

- Ease of Entry
- Supplier Power
- Buyer Power
- Threat of substitutes
- Competitive rivalry

How easy will it be to enter the sector?

This is a complex question and one that breaks down to a number of elements. It has to do with the conservativeness of the market and whether brand loyalty is strong. It has to do with the position of the existing sellers and if there is a dominant monopoly in place. It has to do with the regulatory environment of the industry, whether there are standards and accreditations to be achieved before you can offer your product and it has to do with the use of intellectual property.

Let's take each aspect in turn. Market conservatism is something you will have to judge from your conversations with potential customers. Try to find out who they buy from currently and for how long they have done so. Ask them what it would take for them to switch supplier. Hopefully the answers will have to do with quality, reliability, convenience and the superiority of your solution to their problem. If so then you are likely to have a good competitive advantage. If the answers are all to do with price and brand, then it may be more difficult for you. Competing on price alone is very hard to achieve.

Brand loyalty can be very important in some industries. Customers have built up a level of trust with their suppliers and moving them away, especially to a new unknown brand can be very difficult. It may even be that you would stand a better chance by including an existing supplier within the consortium and exploiting the results through them, rather then competing with them directly. Licensing the technology to existing suppliers is also an option in these circumstances.

Industries that are inhabited by a dominant monopoly are exceedingly difficult to penetrate. The monopoly is likely to strenuously resist newcomers and they will have the power to beat you on price and promotional activity to such an extent that even reaching potential customers could be difficult. Imagine how hard it would be to launch a completely new supermarket chain at the moment.

Most industries are regulated in some way. The vast majority of products require safety and conformity certification and liability insurances. Many services are also regulated with the service provider being required to belong to professional associations which accredit the quality and standards of the service. You will need to understand and conform to all these requirements before you can sell your first product or service and in many cases you will be required to have some quality systems in place even during the design and development stages of the project. The International Organisation for Standardization (ISO) is a good place to start your search of relevant standards that might apply to you.

Finally you need to be sure that you are entitled to provide your product or service without infringing someone else's intellectual property rights. Maintaining an IP log during the project will help to identify background IP that you rely on but you will also need to be sure that what you consider to be your innovations have not already been invented and protected by someone else.

Industry sectors with high barriers to entry are not entirely bad however, because if you have a particular innovation that allows you to enter, it does at least protect you from other new entrants that might want to follow you.

What power do your suppliers have over you?

To answer this question we need to look at all the players that are involved between your taking an order from a customer and completing delivery. This could include suppliers of parts or components, delivery companies, other service providers, sub-contractors, anybody that you rely on in the process of satisfying that order. Make a list of them all.

Now for each one, write down whether what they supply you with is unique or if you could get an equivalent from somewhere else. Any that are unique are of concern– what would you do if they suddenly increased their prices, demanded more cash up front or told you they could not satisfy your demand? Think how you can multi-source as many of your supplies as possible. You may never need to, but it will help to keep your suppliers honest.

What power do your customers have over you?

This is closely related to the previous question but this time we are looking at who sets the terms of sale, you or your customer. In some cases, you will be able to get customers to pay at the time of order or at the point of delivery, but in many business to business transactions you will have to invoice and hope for payment within 30 days or so.

Another aspect of this question is the extent to which your customers are sensitive to price. Obviously no one likes to pay more than necessary but for example if you were to set your prices 10% higher, what effect would that have on your customers decision to buy from you?

One final element to this question is to find out how well informed your customers are to the alternatives and to the market as a whole. Do they continually compare different suppliers or are they content to place long standing orders with you? This will have an effect on their control over your business.

How strong is the threat of people copying your idea?

This question relates to how you are able to protect the foreground IP that the project generates. You will also need to think about how else the competition could solve the customer's problem. Put yourself in the competition's position, how would you react to your new idea and how would you try to compete with it?

What is the competitive environment like within the industry?

This is a difficult question to gauge from the outside but there are some clues that you can look for. Firstly how many companies are in the sector? If the answer is lots then it is likely that they have to compete strongly with each other for business.

How fast is the industry sector growing? As we have seen already, if the sector is growing well, rivalry will be reduced as companies will still be able to grow without stealing market share from the competition.

How much differentiation is there between the existing players and between their plans and yours? The higher the differentiation the more likely it is that everybody is targeting slightly different niches and that competition will be on specification, performance and quality rather than price.

How easy is it for customers to switch suppliers? The easier it is for customers to change, the more rivalry there will be between the suppliers.

Another indicator of rivalry is how much, if any, collaborative work goes on between the different companies in the sector. Do they collaborate in the setting of standards for the industry, is there a vibrant trade association that shares best practice and are there any collaborative research and development programmes that are extending the technology used in the sector? If so the rivalry is probably relatively low and if you are entering the sector with a new advance, you could find that there are license opportunities, even investment interest from the larger players, in addition to direct sales within the market. How easy would it be for you to attract one of the existing players to join your project?

So what do you think of the new industry you are going to expand into? Will it let you compete on the basis of the brilliance and novelty of your project results or is it going to be battle to survive, driven entirely by price? Your response here is just as important as the earlier analysis of the market need for your product. If the answers are worrying, it is very worth while looking for more partners who already have a foot hold in the sector or sticking to your existing sector.

8.5 Exploitation plan

The third activity in the process should take place towards the end of the project as it develops the plan for exploiting the results after the project has finished. The exploitation plan is essentially a business plan that explains all the partners' activities and tasks that are required to see a return on their investment.

Writing the exploitation plan will require a little more thought and analysis and is also likely to benefit from the involvement of the partners' marketing and corporate personnel as well as the technical team that have been working on the project.

Depending on the complexity of the project objectives, you may find that you need to create more than one exploitation plan, one for each of the major market opportunities that the consortium wish to explore. In fact some partners may wish to develop their own plans unilaterally.

The plan need not be a formal document, however you will find that creating a formal document will be highly beneficial as it helps to identify the issues and ensure that crucial aspects are not neglected.

The structure of the exploitation plan is down to you as it will differ from opportunity to opportunity, but in most cases it should at least include the following sections;
- The market opportunity
- The intended route to market
- Additional development requirements
- Standards and accreditation
- The plan for the future

Consider who the audience will be for the document: Will it be read by senior management, funding organisations or potential exploitation partners? The answer will determine how much introduction and background information will be required.

The market opportunity

Set out the market opportunity that you have analysed during the project. Outline the market sector in terms of size, performance and characteristics. Describe the customers you have identified and set out the scale of the opportunity. You can précis your use cases to explain how these customers will benefit from the project results. It might also be worth including the full use cases as an annex.

The route to market

Describe the marketing approach and strategy that will be employed to reach the customers. If the market is very large, describe how you plan to segment it, describe the first few segments you plan to address and justify your choice.

Finally describe the market reach that the project partners have, identify any gaps that exist and explain how they will be filled, either through the expansion of the project team or through licence deals with other organisations in the supply chain.

Additional development requirements

Many collaborative projects, especially those that are grant funded, are pre-competitive and therefore stop short of developing a product or process that can be implemented immediately. Very often there will be further development that needs to be carried out and this needs to be explained in this section of the exploitation plan.

Set out the objectives of the follow-on project that will be required to develop the results to the point of exploitation.

Standards and accreditation

Describe the process by which relevant standards need to be met. For example, new medical devices will need to satisfy various standards and pass safety tests before they can be CE marked in Europe or FDA approved in the US. Provide estimates of the time and effort this will take and who will be responsible for the process.

The plan for the future

In essence this section describes the plan for what happens next. This should be considered and developed in exactly the same way as the original project plan and in fact you can use exactly the same methodology as that described in Chapter 3.

The document should contain a Gantt chart showing the activities or work packages that are involved in taking the project results to market, the risks associated with the process and the resources and time that will be required.

The plan should also describe how the process will be managed and what responsibilities each of the partners will take. Not all the partners may be involved in this stage, for example any academic partners may have contributed all they can by this stage. It is also worth considering whether the consortium agreement needs to be amended or extended to cover the commercialisation activities. Details such as the exact schedule of royalties to be paid may now need to finalised.

Presentation

Depending on the audience, you might also wish to include an executive summary and present the document in a bound fashion as one of the project deliverables. The same rules apply to the presentation of an exploitation plan as to regular business plans. Keep the language clear, include diagrams and pictures to help illustrate the points you are making and present the document in a smart, but not too ostentatious binding.

8.6 Licensing

Sorting out the licensing is a very important part of planning the exploitation of your project results. The more you can agree and write into the consortium agreement at the start of the project the easier it will be at the end.

It is not uncommon for the consortium agreement entered into in order to manage the project is replaced during the exploitation stage with a commercial agreement that defines each partner's access to the project intellectual property and what commercial terms will be used to share the benefits. Licences and royalty payments are the most common mechanisms to achieve these goals.

Access to background intellectual property

Background IP includes all the protected intellectual property that was originally owned by members of the consortium at the start of the project and which will be used in the exploitation of the results. Typically each partner will have brought background IP with them when they joined the consortium and will have allowed the others free access and use of it during the project's development work. This will have been carried out under the terms of the consortium agreement.

At the exploitation stage, access to this IP needs to be re- agreed in order that the results can be exploited commercially by the project partners. Typically this will involve the granting of non-exclusive licenses in return for royalty payments.

The process is made a lot easier if an IP log has been maintained of all the background IP that has been used during the project and will be required for the subsequent exploitation.

Ownership of foreground intellectual property

Foreground IP includes all the innovations and inventions made during the project. The consortium agreement should define who owns and has the right to register and protect these new inventions and how the other partners are allowed access to them during the project.

Again, access to this IP needs to be formally agreed and if necessary the exact nature of royalty payments defined in the new commercial agreement to allow exploitation.

Third party licensing

In some cases you may wish to licence the technology to a third party, someone who was not involved in the project, but who may have access to other markets or territories that the original consortium cannot reach. All the partners would need to agree the terms of such a licence and the share of royalties or other benefits that they could receive.

8.7 Exploitation vehicles

An increasingly common method of exploiting collaborative project results is to create a new company to act as an exploitation vehicle. The foreground IP is assigned to the new company and licences for background IP are provided to allow the company to fully exploit the results.

Ownership of the company is shared between the project partners, either in equal share or by proportion to the value of resources committed by each partner to the project. A shareholders agreement commits each partner to support the new company and its activities.

This approach might be worth considering if the project results in new products or processes that do not immediately fit with any of the existing partners' portfolios. It also protects each of the partners from relying on a single exploitation partner that could switch commercial focus away from the project, lose interest in the opportunity or go into liquidation.

The risks of this approach are that the new company will not have a track record in the market or any brand loyalty with the intended customers. There is also the risk that if the vehicle failed, the IPR would be lost with it. One potential safeguard would be for each partner to provide exclusive licences to the necessary foreground and background IPR to the vehicle company initially and only assign it permanently when commercial targets had been reached.

8.8 Dissemination

Dissemination is a very important part of many collaborative projects, especially when academic partners are involved.

I just wanted to remind you that dissemination needs to be planned carefully as it can have a damaging effect on the protection of new IP. With the exception of the US, patent protection is only possible if the invention has not previously been disclosed to the public. Conference presentations and journal articles count as publication to the public and can prevent patents from being granted. You and your partners would be powerless to prevent anyone from copying your project results and selling them in competition with you anywhere in the world.

Most consortium agreements stipulate that any proposed presentation or publication be shown to all the partners in advance to allow them to edit or veto its release. This should continue to be the case after the project has ended and during the run up to exploitation, at least until the inventions have been protected.

8.9 So in conclusion

1. Exploitation is the process you go through to achieve the desired benefits from the project. It might include
- Launching a new product, process or service
- Entering a new market
- Creating license opportunities
- Publishing papers
- Carrying out further development
- Winning further grants

2. The exploitation methodology consists of three stages:
- Developing a business case for the project
- Developing market understanding
- Creating an exploitation plan

3. Developing a market understanding consists of:
- Analyzing the market opportunity
- Understanding the market dynamics
- Understanding the industry forces

4. Analysing the market opportunity consists of asking the following questions:
- What problem is being solved
- How will customer behaviour change
- Who else is affected

5. Understanding the market dynamics consists of asking the following questions:
- Where are the customers
- How many are there
- How is the market size changing

- What trends affect the business opportunity

6. There are five industry forces:
    - Ease of entry
    - Supplier power
    - Buyer power
    - Threat of substitutes
    - Competitive rivalry

7. The exploitation plan should consist of:
    - The market opportunity
    - The intended route to market
    - Additional development requirements
    - Standards and accreditation issues
    - Plan for the future

8. Licensing issues need to be addressed that impact:
    - Assess to background IPR
    - Ownership of foreground IPR
    - Access by thirst parties

9 Project completion

The final stages of the project are very important and have a disproportional affect on the overall success of the project. Your consortium will need to remain vigilant to any remaining risks and ensure that all outstanding work is completed satisfactorily. This will be an particularly busy period for the project manager especially if grant funding is involved and the funding organisation requires the preparation of detailed final reports and evaluations. Even if this is not the case, there will be internal reviews and lessons to be learned from an analysis of the projects performance.

The partners' focus should shift back to the original objectives for the project. During the execution stages it is easy to forget about the objectives and simply concentrate on delivering the work packages. Now the reasons for carrying out the project must be restated and any additional tasks required to fully meet the objectives will need to be planned and implemented.

The social aspects of the project also need considering. Some of the partners will not be carrying on into the exploitation stages and some members of the team will be disbanding to work on new projects. Some sort of end of project party or dinner is probably in order.

9.1 Reaching the finishing line

As the project draws to a close, the partners will need to pay special attention to outstanding tasks to ensure that they are completed by the end of the project. There are a number of special actions that can be taken to help with this process:

- Produce and review checklists of outstanding work
- Plan the remaining tasks to a finer level of detail to provide tighter control
- Hold more steering committee meetings to ensure problems are identified, understood and solutions actioned without delay
- Creating a task force with special responsibilities for completing outstanding work
- Closing contracts with suppliers and subcontractors to ensure unnecessary costs are contained
- Provide additional support to the project manager

Checklists of the outstanding work should be created at a work package and project level. They should describe the work required and identify who is responsible for finishing it. The steering committee can then review progress and tick off the tasks as they are completed.

By planning the remaining tasks in finer detail, the remaining work is better understood and stands a better chance of being completed by the deadline. If you have not been carrying out a rolling wave planning process so far, then this is the time to start. During the last quarter or two of the project, all remaining work packages should be planned to a resolution of weeks.

Steering committee meetings should be ramped up towards the end of the project. If they are normally held once a quarter, this should increase to twice in the penultimate quarter and could even need to be increased further to monthly during the final quarter. The committee will also be able to review progress of the exploitation plans over the same period.

It may also be worth considering providing the project manager with a deputy to help during this stage of the project. An individual with finishing skills and an eye for detail could be invaluable to help large and complex projects finish successfully.

9.2 Documentation

In cases where the project has resulted in a prototype or demonstrator, its operation and design is likely to require detailed documentation to allow others to operate and understand it. Documentation work tends to use up a lot of time towards the end of R&D projects but is essential for the exploitation of the results. Make sure that you plan sufficient time for these activities and that the quality of reports remains high.

Writing documentation is not everyone's idea of fun and writing skills may vary so it might be worth considering creating a documentation chief or small team that takes basic technical information from the project development teams and creates good quality, consistent reports and instructions.

Grant funding bodies almost always require that a final report is prepared that explains how the project succeeded in meeting the original objectives and how their funding was spent. If this is the case you should find out from the funding body the exact nature of the document they are expecting. Some require detailed reports that will take you many weeks to write while others may provide you with a template or other guidance to help you. It is worth noting that if you leave the preparation of this document until after the official end date of the project, you are unlikely to be able to claim the time it takes to write as an eligible cost of the project.

Final reports are often written in an upbeat way, celebrating the successes of the project but it should remain factual and realistic about the achievements. The lessons learned from the management of the project as well as the technical work should be included.

9.3 Post completion reviews

There are many good reasons for carrying out a review at the end of the project. These include:

- Developing an understanding of what objectives were achieved and of those not achieved, the reasons for the disappointment.
- Developing an understanding of how the final costs differed from the original plan. This could be useful for future planning purposes
- A review of the successes and failures of the project to provide lessons for future projects and activities.

If grant funding was provided for the project there is a very high chance that the funding organisation will require a review of the expenditure and the benefits achieved by the project. This is often contained within a final report but many funding bodies retain the right to re-visit the partners in the years following the project in order to learn how the project provided long term business benefits for the partners. Make sure you understand what is expected in the final reporting so that sufficient resource can be planned to accommodate it.

9.4 Project close meeting & festivities

The project close meeting is in some ways just as important as the launch meeting all those years before. It allows everyone to come together, possibly for the last time and celebrate the successes of the project.

The meeting should review both the positive and negative aspects of the project. Achievements should be celebrated and rewarded and under-achievements understood and possibly chastised. The lessons learned should be discussed and agreed upon as these will not only be carried forward to the exploitation work but go with the individuals to their next projects as the team disbands.

The exploitation plans should also be reviewed and launched so that everybody understands what is to happen with the results, even if they will not personally be involved.

There should also be a social event to mark the end of the project. A dinner might be appropriate or perhaps something more fitting with the project area. You will know the team well enough by now to pick an appropriate celebration. Some large projects even provide small gifts or memorabilia to all the project participants as something to remember the project by. The partners might also want to buy the project manager a bottle of something appropriate to show their gratitude for all their hard work.

9.5 So in conclusion

1. The key requirements for achieving successful project completion are:
- Finishing the work
- Ensuring the outcomes are fully documented
- Reviewing the progress achieved
- Closing the project and disbanding the partnership

2. The work must be finished in a timely and efficient way. The following can help:
- Checklists of outstanding work
- Finer planning of later tasks
- More frequent steering committee meetings
- Creation of task forces
- Efficiently closing sub-contracts
- Supporting the project manager

3. Documentation needs to be completed. This can be aided through nominating a documentation chief. The requirements of any final

report should be understood well in advance to allow time for its preparation before the team disbands.

4. Post completion reviews should be held to:
- Document what has been achieved
- Compare the planned costs with the actual costs
- Generate a set of lessons learned from the project

5. The project close meeting is an important part of the project and should include:
- Positive and negative aspects of the project
- Lessons learned
- Exploitation of the results
- A social event or party

10 Closing remarks

Look back over the ground we have covered in this book. We started with a look at the funding and legal arrangements of collaborative R&D projects. Getting these basics right can make a big difference to the smooth running of the project and in winning financial support.

We looked at building your project on the sound foundations of a strong business case and a robust project plan. Both of these aspects are standard in project management but are a little more complex in the case of collaborative projects. Firstly all the partners have different motives for being involved and these need to be considered carefully to ensure that everyone remains committed to the project. Secondly the partners are often based hundreds of miles apart and so the planning and communications issues need special consideration.

We then looked at the general running of the project to ensure that the momentum is kept and that everyone works well together. Many grant funded projects are monitored externally and we looked at how best to work with monitoring officers and to use their involvement to add value to the project.

Next we looked at the different motivations and working characteristics of academic and industrial partners. These chapters were written to help introduce each type of partner to the other. I had to generalise a little bit, but having worked as both myself, I can get away with it.

We then took a look at what can go wrong with collaborative projects and how to best deal with the special problems they can create. This included the issues created when a partner leaves the project mid way through, which unfortunately is not an uncommon situation.

Next, and perhaps most importantly, we looked at the exploitation of all the hard work. Exploitation is the raison d'être of the collaborative project and a methodology was presented in three stages. The first stage involved the business case which provides the justification for the project. Developing a sound market understanding then helps focus the project towards the best opportunities and finally the exploitation planning stage provides the way forward after the project is complete.

Finally we looked at the special issues that affect the end of the project including strategies for ensuring that the project finishes on time. We looked at post project evaluation and the need for a social event to mark the end of the project.

If you have read this book in preparation for running or joining a collaborative project, I hope it has prepared you for an exciting and rewarding experience. If your project is already running, I hope it has helped you make the best of the effort you are putting in. The rest is up to you. I wish you the best of success, not just for the project itself, but for the benefits you are able to reap from the results.

Glossary

Application
In requesting grant funding, project consortia complete an *application* which describes their project and which is *assessed* by the funding organisation.

Assessment
Describes the process by which a project *application* is selected for grant funding by a *funding body*.

Background Intellectual Property (Background IP)
IP that is previously known and brought to the project by one of the partners to allow the work to proceed.

Baseline
Within a project plan or *spend profile, baseline* refers to the initial forecast of project time and costs at the start of the project. As the project progresses, actual achieved rates can be compared to the Baseline to illustrate whether the project is on schedule.

Claim
Grants are normally paid in arrears against project expenditure. The mechanism for paying grants involves the consortium making a claim for their grant. A claim form may be provided which describes what costs have been incurred and hence what grant is due.

Consortium
The group of project partners that are working collaboratively on the project, are sharing the risks and benefits, and who have signed the *consortium agreement*.

Consortium Agreement
A formal agreement entered into by all the consortium partners. The agreement sets down the terms of the collaboration, the management of the project, the distribution of intellectual property and what happens if a partner defaults or leaves the project. The content of the agreement is normally for the project partners to agree.

Contribution in kind
A contribution to the project made by an organisation outside of the *consortium*. Contributions are normally in the form of cheap materials or access to facilities. The value of contributions can add to the project's *eligible costs*.

Deliverables
Tangible outputs or results from the project. These could include prototypes, reports, demonstrations etc. The balance of internal and external *deliverables* indicates the extent to which the project is *disseminating* its results.

Deliverables – External
External deliverables are those that are shared with the wider community such as a published paper or public demonstration.

Deliverables – Internal
Internal deliverables are private to the *consortium*.

Dissemination
The sharing of project outputs and results within the wider community.

Eligible Costs
Costs accrued by the project partners in carrying out the project that are acceptable to a *funding body* as costs against which a *grant* contribution can be paid. These are mostly agreed in advance of the *offer letter*.

End Date
This date marks the end of the project. Typically in grant funded projects, no costs can be counted as *eligible* after this date.

Evaluation
An analysis of the effectiveness of the project that is carried out shortly after the project has been completed.

Exploitation
Activities that achieve a commercial return or sustainability benefit, for the consortium partners, from the results of the project.

Financial Viability
Financial Viability is a measure of a *partner's* commercial health and ability to participate in a project. Partners must have sufficient cash flow to fund their project activities in advance of any grant funding which is paid in arrears. Viability can also be extended to a partner's ability to fund exploitation of the results after the project has been completed.

Forecast
Within a spend profile, cost forecasts should be provided and updated to show how the project will accrue costs between now and the end of the project. At the start of the project, the forecast will be the same as the *baseline*.

Foreground Intellectual Property (Foreground IP)
New IP that is generated as part of the project.

Funding Body
Organisations that provide *grant* funding for projects.

Grant
The contribution made by the *funding body* to the project. Grants are normally provided as a percentage of the *eligible costs*.

Intellectual Property (IP)
Ideas, know-how or techniques that allow you to solve a problem.

Intellectual Property Rights (IPR)
The legal ownership of *intellectual property*. The rights allow you to exclusively benefit from the IP.

Lead Partner (a.k.a. Project Leader)
Within each *consortium*, one of the partners will act as the lead. This partner is normally responsible for communication with *funding bodies* and provides the *project manager*.

Micro Organisation
Micro organisations are defined as sole proprietorships (a.k.a. Sole trader), partnerships and limited companies with fewer than 10 employees. There are special conditions under which micro Organisations should be considered as *partners*.

Milestones
Milestones are points within the project that mark particular achievements. Milestones need not coincide with a specific *deliverable*.

Monitoring Officer
An individual appointed by a *funding body* to monitor the progress of a project and the spending of *grant* funds.

Offer Letter
The contract between the *funding body* and the project *consortium*. The offer letter sets out the terms under which the *grant* is provided.

Offer Letter – Hub and Spoke
A hub & spoke offer letter is a single document that is addressed to the *lead partner*. In these cases, the claims are made through the *lead partner* who then distributes *grant* payments when they are received.

Offer Letter – Individual
Some projects have individual offer letters, one for each member of the *consortium*. In these cases, each *partner* makes their own *grant* claim directly.

Partner
A member of the project *consortium*.

Project Leader
See lead partner

Project Schedule (a.k.a. Project Plan)
A document that describes the content and organisation of the *work packages* within the project. The plan should also contain a Gantt chart illustrating the timing of the work packages.

Quarterly
Three monthly intervals starting on the project *start date*.

Scope
Scope is defined as the range of activities and deliverables that the project intends to carry out.

SME – Small and Medium Enterprise
SME is an European definition. In some cases SMEs can receive larger *grant* percentages than large companies.

Spend Profile
The spend profile is a spreadsheet that describes the spread of costs for each partner over the duration of the project. It is used to track expenditure and provide forecast information about project expenditure.

Start Date
The start date is the official start of the project. Typically in *grant* funded projects, no *eligible costs* can be accrued before the start Date.

Technology Transfer

Technology transfer is the dissemination or sharing of technological know-how between partners within a consortium to facilitate benefits or commercial opportunities. Technology transfer is commonly between academic institutions and end-user commercial partners, or between technology based *SMEs* and larger market players.

Vire or Virement

Virement is the movement of budget from one cost heading to another or between one *partner* and another.

Index

use case, 132

Wiki, 89
withdrawal, 115

work package, 67, 82, 153
work packages, 153
working days, 15

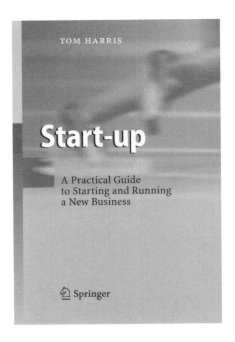

Tom Harris, Lightwater, UK

Start-up

A Practical Guide to Starting and Running a New Business

2006, VIII, 165 p. 24 illus. Hardcover
ISBN-13: 978-3-540-32981-7

So your dream is to start your own business. But how do you know if it will work and how do you make it a reality?

Start-up is a user guide for aspiring entrepreneurs and provides expert advice and guidance on every aspect of launching a new business. It will be of particular value if you are an academic wishing to exploit the commercial value of a new technology or business solution through the creation of a new company. Step-by-step, this inspiring and highly readable book covers how to evaluate the strength of your business idea, how to protect your invention, what legal steps and responsibilities are involved in forming a company, how to position your products in the market, how to create a business plan and raise finance.

The case studies, practical exercises and tips in this book will help to demystify the process of starting a new business, give you the confidence to do it and greatly increase your chances of realising your dream.

Contents: How Good Is Your Idea? – How Can You Protect Your Ideas? – What Is a Company? – How Do You Market Your Product? – How Do You Raise Finance? – How Do You Create a Financial Model? – How Do You Write a Business Plan? – Your Role and Your Team. – What Next?